MOTHS OF
THE LIMBERLOST

BY

GENE STRATTON-PORTER

British Library Cataloguing-in-Publication Data
A catalogue record for this book is available from the
British Library

Contents

Gene Stratton Porter

Gene Stratton Porter was born Geneva Grace Stratton was born in rural Indiana, USA in 1863. Her mother died when she was very young, and after her family moved to the nearby Wabash city, Gene suffered a skull fracture. While recovering, she met pharmacist Charles Darwin Porter, who she married in 1886.

Porter settled in Geneva, Indiana, near Limberlost Swamp, and immersed herself in her nature studies. She slowly moved from sketches and photographs onto poems and short stories, and during the 1890s began to publish her work in magazines such as *Metropolitan, Recreation*, and *Outing*. In 1903, she published her first novel, *Song of the Cardinal.*

The following year, Porter published her best-remembered work, *Freckles* (1904). Set in the wooded wetlands and swamps of central Indiana, the novel was a huge success, as was Porter's next work, *A Girl of the Limberlost* (1909). Between them, the books were estimated to have had more than 40 million readers, and were translated into many languages. *A Girl of the Limberlost* has been adapted for film on four separate occasions.

Over the course of her career, Porter published more than twenty books, many of them works of natural history.

Some of her most popular were *At the Foot of the Rainbow* (1907), *What I Have Done with Birds* (1907), *The Harvester* (1911), *Laddie* (1913), *Michael O'Halloran* (1915), and *A Daughter of the Land* (1918).

In 1920, Porter moved to Los Angeles, California. Her last novel, *Her Father's Daughter* (1921), presented a unique window into her feelings about World War I-era racism and nativism. Porter died in 1924 following a streetcar accident, aged 61.

MOTHS OF THE LIMBERLOST
A book about Limberlost Cabin

by

Gene Stratton-Porter

To
Neltje Degraff Doubleday

"All diamonded with panes of quaint device,
Innumerable of stains, and splendid dyes,
As are the Tiger Moth's deep damask wings."

CHAPTER I.
MOTHS OF THE LIMBERLOST

To me the Limberlost is a word with which to conjure; a spot wherein to revel. The swamp lies in north-eastern Indiana, nearly one hundred miles south of the Michigan line and ten west of the Ohio. In its day it covered a large area. When I arrived; there were miles of unbroken forest, lakes provided with boats for navigation, streams of running water, the roads around the edges corduroy, made by felling and sinking large trees in the muck. Then the Winter Swamp had all the lacy exquisite beauty of such locations when snow and frost draped, while from May until October it was practically tropical jungle. From it I have sent to scientists flowers and vines not then classified and illustrated in our botanies.

It was a piece of forethought to work unceasingly at that time, for soon commerce attacked the swamp and began its usual process of devastation. Canadian lumbermen came seeking tall straight timber for ship masts and tough heavy trees for beams. Grand Rapids followed and stripped the forest of hard wood for fine furniture, and through my experience with the lumber men "Freckles'" story was written. Afterward hoop and stave men and local mills took the best of the soft wood. Then a ditch, in reality a canal, was dredged across the north end through, my best territory, and that carried the water to the Wabash River until oil men could enter the swamp. From that time the wealth they drew to the surface constantly materialized in macadamized roads, cosy homes, and big farms of unsurpassed richness, suitable for growing onions, celery, sugar beets, corn and potatoes, as repeatedly has been explained in everything I have written of the place. Now, the Limberlost exists only in ragged spots and patches, but so rich was it in the beginning that there is yet a wealth of work for a lifetime remaining to me in these, and river thickets. I ask no better hunting grounds for birds, moths, and flowers. The fine roads are a convenience, and settled farms a protection, to be taken into consideration, when bewailing its dismantling.

It is quite true that "One man's meat is another's poison." When poor Limber, lost and starving in the fastnesses of the swamp, gave to it a name, afterward to be on the lips of millions; to him it was deadly poison. To me it has been of unspeakable interest, unceasing work of joyous nature, and meat in full measure, with occasional sweetbreads by way of a treat.

Primarily, I went to the swamp to study and reproduce the birds. I never thought they could have a rival in my heart. But these fragile night wanderers, these moonflowers of June's darkness, literally "thrust themselves upon me." When my cameras were placed before the home of a pair of birds, the bushes parted to admit light, and clinging to them I found a creature, often having the bird's sweep of wing, of colour pale green with decorations of lavender and yellow or running the gamut from palest tans darkest browns, with markings, of pink or dozens of other irresistible combinations of colour, the feathered folk found a competitor that often outdistanced them in my affections, for I am captivated easily by colour, and beauty of form.

At first, these moths made studies of exquisite beauty, I merely stopped a few seconds to reproduce them, before proceeding with my work. Soon I found myself filling the waiting time, when birds were slow in coming before the cameras, when clouds obscured the light too much for fast exposures, or on grey days, by searching for moths. Then in collecting abandoned nests, cocoons were found on limbs, inside stumps, among leaves when gathering nuts, or queer shining pupae-cases came to light as I lifted wild flowers in the fall. All these were carried to my little conservatory, placed in as natural conditions as possible, and studies were made from the moths that emerged the following spring. I am not sure but that "Moths of Limberlost Cabin" would be the most appropriate title for this book.

Sometimes, before I had finished with them, they paired, mated, and dotted everything with fertile eggs, from which tiny caterpillars soon would emerge. It became a matter of intense interest to provide their natural foods and raise them. That started me to watching for caterpillars and eggs out of doors, and friends of my work began carrying them to me. Repeatedly, I have gone through the entire life process, from mating newly emerged moths, the egg period, caterpillar life, with its complicated moults and changes, the spinning of the cocoons, the miraculous winter sleep, to the spring appearance; and with my cameras recorded each stage of development. Then on platinum paper, printed so lightly from these negatives as to give only an exact reproduction of forms, and with water colour medium copied each mark, line and colour gradation in most cases from the living moth at its prime. Never was the study of birds so interesting.

The illustration of every moth book I ever have seen, that attempted coloured reproduction, proved by the shrivelled bodies and unnatural position of the wings, that it had been painted from objects mounted from weeks to years in private collections or museums. A lifeless moth fades rapidly under the most favourable conditions. A moth at eight days of age, in the last stages of decline, is from four to six distinct shades lighter in colour than at six hours from the cocoon, when it is dry, and ready for flight. As soon as circulation stops, and the life juices evaporate from the wings and body, the colour grows many shades paler. If exposed to light, moths soon fade almost beyond recognition.

I make no claim to being an entomologist; I quite agree with the "Autocrat of the Breakfast Table", that "the subject is too vast for any single human intelligence to grasp." If my life depended upon it I could not give the scientific name of every least organ and nerve of a moth, and as for wrestling with the thousands of tiny species of day and night or even attempting all the ramifications of—say the alluringly beautiful Catocalae family—life is too short, unless devoted to this purpose alone. But if I frankly confess my limitations, and offer the book to my nature-loving friends merely as an introduction to the most exquisite creation of the swamp; and the outside history, as it were, of the evolution of these creatures from moth to moth again, surely no one can feel defrauded. Since the publication of "A Girl of the Limberlost", I have received hundreds of letters asking me to write of my experiences with the lepidoptera of the swamp. This book professes to be nothing more.

Because so many enemies prey upon the large night moths in all stages, they are nowhere sufficiently numerous to be pests, or common enough to be given local names, as have the birds. I have been compelled to use their scientific names to assist in identification, and at times I have had to resort to technical terms, because there were no other. Frequently I have written of them under the names by which I knew them in childhood, or that we of Limberlost Cabin have bestowed upon them.

There is a wide gulf between a Naturalist and a Nature Lover. A Naturalist devotes his life to delving into stiff scientific problems concerning everything in nature from her greatest to her most minute forms. A Nature Lover works at any occupation and finds recreation in being out of doors and appreciating the common things of life as they appeal to his senses.

The Naturalist always begins at the beginning and traces family, sub-family, genus and species. He deals in Latin and Greek terms of resounding and disheartening combinations. At his hands anatomy and markings become lost in a scientific jargon of patagia, jugum, discocellulars, phagocytes, and so on to the end of the volume. For one who would be a Naturalist, a rare specimen indeed, there are many volumes on the market. The list of pioneer lepidopterists begins authoritatively with Linnaeus and since his time you can make your selection from the works of Druce, Grote, Strecker, Boisduval, Robinson, Smith, Butler, Fernald, Beutenmuller, Hicks, Rothschild, Hampson, Stretch, Lyman, or any of a dozen others. Possessing such an imposing array of names there should be no necessity to add to them. These men have impaled moths and dissected, magnified and located brain, heart and nerves. After finishing the interior they have given to the most minute exterior organ from two to three inches of Latin name. From them we learn that it requires a coxa, trochanter, femur, tibia, tarsus, ungues, pulvillus, and anterior, medial and posterior spurs to provide a leg for a moth. I dislike to weaken my argument that more work along these lines is not required, by recording that after all this, no one seems to have located the ears definitely. Some believe hearing lies in the antennae. Hicks has made an especial study of a fluid filled cavity closed by a membrane that he thinks he has demonstrated to be the seat of hearing. Leydig, Gerstaecker, and others believe this same organ to be olfactory. Perhaps, after all, there is room for only one more doctor of science who will permanently settle this and a few other vexing questions for us.

But what of the millions of Nature Lovers, who each year snatch only a brief time afield, for rest and recreation? What of the masses of men and women whose daily application to the work of life makes vacation study a burden, or whose business has so broken the habit of study that concentration is distasteful if not impossible? These people number in the ratio of a million to one Naturalist. They would be delighted to learn the simplest name possible for the creatures they or their friends find afield, and the markings, habits, and characteristics by which they can be identified. They do not care in the least for species and minute detail concerning anatomy, couched in resounding Latin and Greek terms they cannot possibly remember.

I never have seen or heard of any person who on being shown any one of ten of our most beautiful moths, did not consider and promptly pronounce it the most exquisite creation he ever had seen, and evince a lively interest in its history. But when he found it necessary to purchase a text-book, devoid of all human interest or literary possibility, and wade through pages of scientific dissertation, all the time having the feeling that perhaps through his lack of experience his identification was not aright, he usually preferred to remain in ignorance. It is in the belief that all Nature Lovers, afield for entertainment or instruction, will be thankful for a simplification of any method now existing for becoming acquainted with moths, that this book is written and illustrated.

In gathering the material used I think it is quite true that I have lost as many good subjects as I have secured, in my efforts to follow the teachings of scientific writers. My complaint against them is that they neglect essential detail and are not always rightly informed. They confuse one with a flood of scientific terms describing minute anatomical parts and fail to explain the simple yet absolutely essential points over which an amateur has trouble, wheat often only a few words would suffice.

For example, any one of half a dozen writers tells us that when a caterpillar finishes eating and is ready to go into winter quarters it crawls rapidly around for a time, empties the intestines, and transformation takes place. Why do not some of them explain further that a caterpillar of, say, six inches in length will shrink to THREE, its skin become loosened, the horns drop limp, and the creature appear dead and disintegrating? Because no one mentioned these things, I concluded that the first caterpillar I found in this state was lost to me and threw it away. A few words would have saved the complete history of a beautiful moth, to secure which no second opportunity was presented for five years.

Several works I consulted united in the simple statement that certain caterpillars pupate in the ground.

In Packard's "Guide", you will find this—"Lepidopterous pupae should be...kept moist in mould until the image appears." I followed this direction, even taking the precaution to bake the earth used, because I was very anxious about some rare moths. When they failed to emerge in season I dug them out, only to find that those not moulded had been held fast by the damp, packed earth, and all were ruined. I learned by investigation that pupation takes place in a hole worked out by the caterpillar, so earth must touch these cases only as they lie upon it. The one word 'hole' would have saved all those moths for me.

One writer stated that the tongue cases of some pupae turn over and fasten on the back between the wing shields, and others were strangely silent on the subject. So for ten months I kept some cases lying on their backs with the feet up and photographed them in that position. I had to discover for myself that caterpillars that pupate in the ground change to the moth form with the feet and legs folded around the under side of the thorax, the wings wrap over them, and the tongue case bends UNDER and is fastened between the wings.

For years I could find nothing on the subject of how a moth from a burrowing caterpillar made its appearance. In two recent works I find the statement that the pupa cases come to the surface before the moths leave them, but how the operation is performed is not described or explained. Pupa cases from earth consist of two principal parts: the blunt head and thorax covering, and the ringed abdominal sections. With many feeders there is a long, fragile tongue shield. The head is rounded and immovable of its own volition. The abdominal part is in rings that can be turned and twisted; on the tip are two tiny, needlesharp points, and on each of three rings of the abdominal shield there are in many cases a pair of tiny hooks, very slight projections, yet enough to be of use. Some lepidopterists think the pupa works head first to the surface, pushing with the abdomen. To me this seems impossible. The more one forced the blunt head against the earth the closer it would pack, and the delicate tongue shield surely would break. There is no projection on the head that would loosen or lift the earth.

One prominent lepidopterist I know, believes the moth emerges underground, and works its way to the surface as it fights to escape a cocoon. I consider this an utter impossibility. Remember the earth-encrusted cicada cases you have seen clinging to the trunks of trees, after the insect has reached the surface and abandoned them. Think what would happen to the delicate moth head, wings, and downy covering! I am willing to wager all I possess, that no lepidopterist, or any amateur, ever found a freshly emerged moth from an underground case with the faintest trace of soil on its head or feet, or a particle of down missing; as there unquestionably must be, if it forced its way to freedom through the damp spring earth with its mouth and feet.

The point was settled for me when, while working in my garden, one came through the surface within a few inches of my fingers, working with the tip of the abdomen. It turned, twisted, dug away the dirt, fastened the abdominal tip, pulled up the head, and then bored with the tip again. Later I saw several others emerge in the same way, and then made some experiments that forever convinced me that this is the only manner in which ground pupae possibly could emerge.

One writer I had reason to suppose standard authority stated that caterpillars from Citheronia Regalis eggs emerged in sixteen days. So I boxed some eggs deposited on the eleventh, labelled them due to produce caterpillars on the twenty-seventh and put away the box to be attended on that date. Having occasion to move it on the twentyfourth, I peeped in and found half my caterpillars out and starved, proving that they had been hatched at least thirty-six hours or longer; half the others so feeble they soon became inactive, and the remainder survived and pupated. But if the time specified had been allowed to elapse, every caterpillar would have starved.

One of the books I read preparatory to doing this work asserts concerning spinners: "Most caterpillars make some sort of cocoon or shelter, which may be of pure silk neatly wound, or of silk mixed with hair and all manner of external things—such as pieces of leaf, bark, moss, and lichen, and even grains of earth."

I have had caterpillars spin by the hundred, in boxes containing most of these things, have gathered outdoor cocoons by the peck, and microscopically examined dozens of them, and with the exception of leaf, twig, bark, or some other foundation against which it was spun, I never have seen a cocoon with shred, filament, or particle of anything used in its composition that was not drawn from the spinning tube or internal organism of the caterpillar, with the possible exception of a few hairs from the tubercles. I have been told by other workers that they have had captive caterpillars use earth and excrement in their cocoons.

This same work, in an article on protective colouration, lays emphasis on the statement that among pupa cases artificially fastened to different objects out of doors, "the elimination was ninety-two per cent on fences where pupae were conspicuous, as against fifty-two per cent among nettles, where they were inconspicuous." This statement is elaborated and commented upon as making a strong point for colourative protection through inconspicuousness.

Personally, I think the nettles did the work, regardless of colour. I have learned to much experience afield that a patch of nettles or thistles afford splendid protection to any form of life that can survive them. I have seen insects and nesting birds find a safety in their shelter, unknown to their kind that home elsewhere. The test is not fair enough to be worth consideration. If these same pupae had been as conspicuously placed as on the fence, on any EDIBLE GROWTH, in the same location as the fence, and then left to the mercy of playing children, grazing stock, field mice, snakes, bats, birds, insects and parasites, the story of what happened to them would have been different. I doubt very seriously if it would have proved the point those lepidopterists started out to make in these conditions, which are the only fair ones under which such an experiment could be made.

Many people mentioned in connexion with the specimens they brought me have been more than kind in helping to collect the material this volume contains; but its publication scarcely would have been possible to me had it not been for the enthusiasm of one girl who prefers not to be mentioned and the work of a seventeen-year-old boy, Raymond Miller. He has been my sole helper in many difficult days of field work among the birds, and for the moths his interest reached such a pitch that he spent many hours afield in search of eggs, caterpillars, cocoons, and moths, when my work confined me to the cabin. He has carried to me many of my rarest cocoons, and found in their native haunts several moths needed to complete the book. It is to be hoped that these wonderful days afield have brought their own compensation, for kindness such as his I never can reward adequately. The book proves my indebtedness to the Deacon and to Molly-Cotton. I also owe thanks to Bob Burdette Black, the oldest and warmest friend of my bird work, for many fine moths and cocoons, and to Professor R. R. Rowley for the laborious task of scientifically criticizing this book and with unparalleled kindness lending a helping hand where an amateur stumbled.

CHAPTER II.
MOTHS, EGGS, CATERPILLARS, WINTER QUARTERS

If you are too fastidious to read this chapter, it will be your permanent loss, for it contains the life history, the evolution of one of the most amazingly complicated and delicately beautiful creatures in existence. There are moths that come into the world, accomplish the functions that perpetuate their kind, and go out, without having taken any nourishment. There are others that feed and live for a season. Some fly in the morning, others in the glare of noon, more in the evening, and the most important class of big, exquisitely lovely ones only at night. This explains why so many people never have seen them, and it is a great pity, for the nocturnal, non-feeding moths are birdlike in size, flower-like in rare and complicated colouring, and of downy, silent wing.

The moths that fly by day and feed are of the Sphinginae group, Celeus and Carolina, or Choerocampinae, which includes the exquisite Deilephila Lineata, and its cousins; also Sphingidae, which cover the clear-winged Hemaris diffinis and Thysbe. Among those that fly at night only and take no food are the members of what is called the Attacine group, comprising our largest and commonest moth, Cecropia;

also its near relative Gloveri, smaller than Cecropia and of lovely rosy wine-colour; Angulifera, the male greyish brown, the female yellowish red; Promethea, the male resembling a monster Mourning Cloak butterfly and the female bearing exquisite red-wine flushings; Cynthia, beautiful in shades of olive green, sprinkled with black, crossed by bands of pinkish lilac and bearing crescents partly yellow, the remainder transparent. There are also the deep yellow Io, pale blue-green Luna, and Polyphemus, brown with pink bands of the Saturniidae; and light yellow, red-brown and grey Regalis, and lavender and yellow Imperialis of the Ceratocampidae, and their relatives. Modest and lovely Modesta belongs with the Smerinthinae group; and there are others, feeders and non-feeders, forming a list too long to incorporate, for I have not mentioned the Catocalae family, the fore-wings of which resemble those of several members of the Sphinginae, in colour, and when they take flight, the back ones flash out colours that run the gamut from palest to deepest reds, yellows, and browns, crossed by wide circling bands of black; with these, occasionally the black so predominates that it appears as if the wing were black and the bands of other colour. All of them are so exquisitely beautiful that neither the most exacting descriptions, nor photographs from life, nor water colours faithfully copied from living subjects can do them justice. They must be seen alive, newly emerged, down intact, colours at their most brilliant shadings, to be appreciated fully. With the exception of feeding or refraining from eating, the life processes of all these are very similar.

Moths are divided into three parts, the head, thorax, and abdomen, with the different organs of each. The head carries the source of sight, scent, and the mouth parts, if the moth feeds, while the location of the ears is not yet settled definitely. Some scientists place hearing in the antennae, others in a little organ on each side the base of the abdomen. Packard writes: "The eyes are large and globose and vary in the distance apart in different families": but fails to tell what I want to know most: the range and sharpness of their vision. Another writer states that the eyes are so incomplete in development that a moth only can distinguish light from darkness and cannot discern your approach at over five feet.

This accords with my experience with Cecropia, Polyphemus, Regalis, and Imperialis. Luna either can see better, hear acutely, or is naturally of more active habit. It is difficult to capture by hand in daytime; and Promethea acts as if its vision were even clearer. This may be the case, as it flies earlier in the day than any of the others named, being almost impossible to take by hand unless it is bound to a given spot by sex attraction. Unquestionably the day fliers that feed—the Sphinginae and Choerocampinae groups—have fairly good vision, as also the little "Clear-wings" tribe, for they fly straight to the nectar-giving flowers and fruits they like best to feed upon, and it is extra good luck if you capture one by hand or even with a net. It must be remembered that all of them see and go to a bright light at night from long distances.

Holland writes: "The eyes of moths are often greatly developed," but makes no definite statements as to their range of vision, until he reaches the Catocalae family, of which he records: "The hind wings are, however, most brilliantly coloured. In some species they are banded with pink, in others with crimson; still others have markings of yellow, orange, or snowy white on a background of jet black. These colours are distinctive of the species to a greater or less extent. They are only displayed at night. The conclusion is irresistibly forced upon us that the eyes of these creatures are capable of discriminating these colours in the darkness. We cannot do it. No human eye in the blackness of the night can distinguish red from orange or crimson from yellow. The human eye is the greatest of all anatomical marvels, and the most wonderful piece of animal mechanism in the world, but not all of power is lodged within it. There are other allied mechanisms which have the power of responding to certain forms of radiant energy to a degree which it does not possess."

This conclusion is not "irresistibly forced" upon me. I do believe, know in fact, that all day-flying, feeding moths have keener sight and longer range of vision than non-feeders; but I do not believe the differing branches of the

Catocalae group, or moths of any family, locate each other "in the blackness of night," by seeing markings distinctly. I can think of no proof that moths, butterflies or any insects recognize or appreciate colour. Male moths mate with females of their kind distinctly different from them in colour, and male butterflies pair with albinos of their species, when these differ widely from the usual colouring.

A few moths are also provided with small simple eyes called ocelli; these are placed on top of the head and are so covered with down they cannot be distinguished save by experts. Mueller believes that these are for the perception of objects close to a moth while the compound eyes see farther, but he does not prove it.

If the moth does not feed, the mouth parts are scarcely developed. If a feeder, it has a long tongue that can be coiled in a cleft in the face between the palpi, which Packard thinks were originally the feelers. This tongue is formed of two grooved parts so fastened together as to make a tube through which it takes flower and fruit nectar and the juices of decaying animal matter.

What are thought by some to be small organs of touch lie on either side the face, but the exact use of these is yet under discussion, It is wofully difficult to learn some of these things.

In my experience the antennae, are the most sensitive, and therefore the most important organs of the head—to me. In the Attacine group these stand out like delicately cut tiny fern fronds or feathers, always being broader and more prominent on the male. Other families are very similar and again they differ widely. You will find moths having pointed hair-like antennae; others heaviest at the tip in club shape, or they may be of even proportion but flat, or round, or a feathered shaft so fine as to be unnoticed as it lies pressed against the face. Some writers say the antennae are the seat of scent, touch, and hearing. I had not thought nature so impoverished in evolving her forms as to overwork one delicate little organ for three distinct purposes. The antennae are situated close where the nose is, in almost every form of life, and I would prefer to believe that they are the organs of scent and feeling. I know a moth suffers most over any injury to them; but one takes flight no quicker or more precipitately at a touch on the antennae than on the head, wing, leg, or abdomen.

We are safe in laying down a law that antennae are homologous organs and used for identical purposes on all forms of life carrying them. The short antennae of grasshoppers appear to be organs of scent. The long hair-fine ones of katydids and crickets may be also, but repeatedly I have seen these used to explore the way ahead over leaves and limbs, the insect feeling its path and stepping where a touch assures it there is safe footing. Katydids, crickets, and grasshoppers all have antennae, and all of these have ears definitely located; hence their feelers are not for auricular purposes. According to my logic those of the moth cannot be either. I am quite sure that primarily they serve the purpose of a nose, as they are too short in most cases to be of much use as 'feelers,' although that is undoubtedly their secondary office. If this be true, it explains the larger organs ofthe male. The female emerges from winter quarters so weighted with carrying from two to six hundred eggs, that she usually remains and develops where she is. This throws the business of finding her location on the male. He is compelled to take wing and hunt until he discovers her; hence his need of more acute sense of scent and touch. The organ that is used most is the one that develops in the evolution of any form of life.

I can well believe that the antennae are most important to a moth, for a broken one means a spoiled study for me. It starts the moth tremulously shivering, aimlessly beating, crazy, in fact, and there is no hope of it posing for a picture. Doctor Clemens records that Cecropia could neither, walk nor fly, but wheeled in a senseless, manner when deprived of its antennae. This makes me sure that they are the seat of highest sensibility, for I have known in one or two cases of chloroformed moths reviving and without struggle or apparent discomfort, depositing eggs in a circle around them, while impaled to a setting board with a pin thrust through the thorax where it of necessity must have passed through or very close the nervous cord and heart.

The moth is covered completely with silken down like tiny scales, coloured and marked according to species, and so lightly attached that it adheres to the cocoon on emergence and clings to the fingers at the lightest touch. From the examination of specimens I have taken that had disfigured themselves, it appears that a moth rubbed bare of down would seem as if covered with thinly cut, highly polished horn, fastened together in divisions. This is called 'chitine' by scientists.

The thorax bears four wings, and six legs, each having five joints and ending in tiny claws. The wings are many-veined membranous sacs, covered with scales that are coloured according to species and arranged to form characteristic family markings. They are a framework usually of twelve hollow tubes or veins that are so connected with the respiratory organs as to be pneumatic. These tubes support double membranes covered above and below with down. At the bases of the wings lie their nerves. The fore-wings each have a heavy rib running from the base and gradually decreasing to the tip. This is called the costa. Its purpose is to bear the brunt of air-pressure in flight. On account of being compelled to fly so much more than the females, the back wings of the males of many species have developed a secondary rib that fits under and supports the front, also causing both to work together with the same impulse to flight. A stiff bunch of bristles serves the same purpose in most females, while some have a lobe extending from the fore-wing. As long as the costa remains unbroken to preserve balance, a moth that has become entangled in bushes or suffered rough treatment from birds can fly with badly damaged wing surfaces.

In some species, notably the Attacine group and all non-feeding, night-flying moths, the legs are short, closely covered with long down of the most delicate colours of the moth, and sometimes decorated with different shades. Luna has beautiful lavender legs, Imperialis yellow, and Regalis red-brown. The day-flying, feeding group have longer, slenderer legs, covered with shorter down, and carry more elaborate markings. This provision is to enable them to cling firmly to flower or twig while feeding, to help them to lift the body higher, and walk dextrously in searching for food. It is also noticeable that these moths have, for their size, comparatively much longer, slenderer wings than the non-feeders, and they can turn them back and fold them together in the fly position, thus enabling them to force their way into nectar-bearing flowers of trumpet shape.

The abdomen is velvet soft to the touch, and divided into rings called segments, these being so joined that this member can be turned and twisted at will. In all cases the last ring contains the sex organs. The large abdomen of the female carries several hundred embryo eggs, and that of the male the seminal fluid.

Much has been written of moths being able to produce odours that attract the sexes, and that are so objectionable as to protect them from birds, mice, and bats. Some believe there are scent glands in a few species under the wing scales. I have critically examined scores of wings as to colour markings, but never noticed or smelled these. On some, tufts of bristlelike hairs can be thrust out, that give a discernible odour; but that this carries any distance or is a large factor in attracting the sexes I do not believe so firmly, after years of practical experience, as I did in the days when I had most of my moth history from books. I have seen this theory confounded so often in practice.

In June of 1911, close six o'clock in the evening, I sat on the front veranda of the Cabin, in company with my family, and watched three moths sail past us and around the corner, before I remembered that on the screen of the music-room window to the east there was a solitary female Promethea moth, that day emerged from a cocoon sent me by Professor Rowley. I hurried to the room and found five male moths fluttering before the screen or clinging to the wild grape and sweet brier vines covering it. I opened the adjoining window and picked up three of the handsomest with my fingers, placing them inside the screen. Then I returned to the veranda.

Moths kept coming. We began studying the conditions. The female had emerged in the diningroom on the west side of the cabin. On account of the intense heat of the afternoon sun, that side of the building had been tightly closed all day. At four o'clock the moth was placed on the east window, because it was sheltered with vines. How soon the first male found her, I do not know. There was quite a stiff evening breeze blowing from the west, so that any odour from her would have been carried on east. We sat there and watched and counted six more moths, every one of which came down wind from the west, flying high, above the treetops in fact, and from the direction of a little tree-filled plot called Studabaker's woods. Some of them we could distinguish almost a block away coming straight toward the Cabin, and sailing around the eastern corner with the precision of hounds on a hot trail. How they knew, the Almighty knows; I do not pretend to; but that there was odour distilled by that one female, practically imperceptible to us (she merely smelled like a moth), yet of such strength as to penetrate screen, vines, and roses and reach her kind a block away, against considerable breeze, I never shall believe.

The fact is, that moths smell like other moths of the same species, and within a reasonable radius they undoubtedly attract each other. In the same manner birds carry a birdlike odour, and snakes, frogs, fish, bees, and all animals have a scent peculiar to themselves. No dog mistakes the odour of a cat for that of another dog. A cow does not follow the scent of horses to find other cattle. No moth hunts a dragon-fly, a butterfly, or in my experience, even a moth of another species in its search for a mate. How male moths work the miracles I have seen them accomplish in locating females, I cannot explain. As the result of acts we see them perform, we credit some forms of life with much keener scent than others, and many with having the power more highly developed than people. The only standard by which we can determine the effect that the odour of one insect, bird, or animal has upon another is by the effect it has upon us. That a male moth can smell a female a block away, against the wind, when I can detect only a faint musky odour within a foot of her, I do not credit.

Primarily the business of moths is to meet, mate, and deposit eggs that will produce more moths. This is all of life with those that do not take food. That they add the completing touch and most beautiful form of life to a few exquisite May and June nights is their extra good fortune, not any part of the affair of living. With moths that feed and live after reproduction, mating and egg placing comes first. In all cases the rule is much, the same. The moths emerge, dry their wings, and reach full development the first day. In freedom, the females being weighted with eggs seldom attempt to fly. They remain where they are, thrust out the egg placer from the last ring of the abdomen and wait. By ten o'clock the males, in such numbers as to amaze a watcher, find them and remain until almost morning.

Broad antennae, slenderer abdomen, and the claspers used in holding the female in mating, smaller wings and more brilliant markings are the signs by which the male can be told in most cases. In several of the Attacine group, notably Promethea, the male and female differ widely in markings and colour. Among the other non-feeders the difference is slight. The male Regalis has the longest, most gracefully curved abdomen and the most prominent claspers of any moth I ever examined; but the antennae are so delicate and closely pressed against the face most of the time as to be concealed until especially examined. I have noticed that among the moths bearing large, outstanding antennae, the claspers are less prominent than with those having small, inconspicuous head parts. A fine pair of antennae, carried forward as by a big, fully developed Cecropia, are as ornamental to the moth as splendidly branching antlers are to the head of a deer.

The female now begins egg placing. This requires time, as one of these big night moths deposits from three hundred and fifty to over six hundred eggs. These lie in embryonic state in the abdomen of the female. At her maturity they ripen rapidly. When they are ready to deposit, she is forced to place them whether she has mated or not. In case a mate has found her, a small pouch near the end of her abdomen is filled with a fluid that touches each egg in passing and renders it fertile. The eggs differ with species and are placed according to family characteristics. They may be pure white, pearl-coloured, grey, greenish, or yellow. There are round, flat, and oblong eggs. These are placed differently in freedom and captivity. A moth in a natural location glues her eggs, often one at a time, on the under or upper side of leaves. Sometimes she dots several in a row, or again makes a number of rows, like a little beaded mat. One authority I have consulted states that "The eggs are always laid by the female in a state of freedom upon the food-plant which is most congenial to the larvae." This has not 'always' been the case in my experience. I have found eggs on stone walls, boards, fences, outbuildings, and on the bark of dead trees and stumps as well as living, even on the ground. This also, has been the case with the women who wrote "Caterpillars and their Moths", the most invaluable work on the subject ever compiled.

A captive moth feels and resents her limitations. I cannot force one to mate even in a large box. I must free her in the conservatory, in a room, or put her on an outside window br door screen. Under these conditions one will place her eggs more nearly as in freedom; but this makes them difficult to find and preserve. Placed in a box and forced by nature to deposit her eggs, as a rule, she will remain in one spot and heap them up until she is forced to move to make room for more. One big female Regalis of the last chapter of this book placed them a thimbleful at a time; but the little caterpillars came rolling out in all directions when due. In my experience, they finish in four or five nights, although I have read of moths having lived and placed eggs for ten, some species being said to have deposited over a thousand. Seven days is usually the limit of life for these big night moths with me; they merely grow inactive and sluggish until the very last, when almost invariably they are seized with a muscular attack, in which they beat themselves to rags and fringes, as if resisting the overcoming lethargy. It is because of this that I have been forced to resort to the gasoline bottle a few times when I found it impossible to paint from the living moth; but I do not put one to sleep unless I am compelled.

I never have been able to induce a female to mate after confinement had driven her to begin depositing her eggs, not even under the most favourable conditions I could offer, although others record that they have been so fortunate. Repeatedly I have experimented with males and females of different species, but with no success. I have not seem a polygamous moth; but have read of experiences with them.

Sometimes the eggs have a smooth surface, again they may be ridged or like hammered brass or silver. The shells are very thin and break easily. At one side a place can be detected where the fertilizing fluid enters. The coming caterpillar begins to develop at once and emerges in from six to thirty days, with the exception of a few eggs placed in the fall that produce during the following spring. The length of the egg period differs with species and somewhat with the same moths, according to suitable or unfavourable placing, and climatic conditions. Do not accept the experience of any one if you have eggs you very much desire to be productive of the caterpillars of rare moths; after six days take a peep every day if you would be on the safe side. With many species the shells are transparent, and for the last few days before emergence the growth of the little caterpillars can be watched through them.

When matured they break or eat a hole in their shells and emerge, seeming much too large for the space they occupied. Family characteristics show at once. Many of them immediately turn and eat their shells as if starving; others are more deliberate. Some grace around for a time as if exercising and then return and eat their shells; others walk briskly away and do not dine on shell for the first meal. Usually all of them rest close twenty-four hours before beginning on leaves. Once they commence feeding in favourable conditions they eat enormously and grow so rapidly they soon become too large for their skins to hold them another instant; so they pause and stop eating for a day or two while new skin forms. Then the old is discarded and eaten for a first meal, with the exception of the face covering. At the same time the outer skin is cast the intestinal lining is thrown off, and practically a new caterpillar, often bearing different markings, begins to feed again.

These moults occur from four to six times in the development of the caterpillar; at each it emerges larger, brighter, often with other changes of colour, and eats more voraciously as it grows. With me, in handling caterpillars about which I am anxious, their moulting time is critical. I lost many until I learned to clean their boxes thoroughly the instant they stopped eating and leave them alone until they exhibited hunger signs again. They eat greedily of the leaves preferred by each species, doing best when the foliage is washed and drops of water left for them to drink as they would find dew and rain out of doors. Professor Thomson, of the chair of Natural History of the University of Aberdeen, makes this statement in his "Biology of the Seasons", "Another feature in the life of caterpillars is their enormous appetite. Some of them seem never to stop eating, and a species of Polyphemus is said to eat eighty-six thousand times its own weight in a day." I notice Doctor Thomson does not say that he knows this, but uses the convenient phrase, "it is said." This is an utter impossibility. The skin of no living creature will contain eighty-six thousand times its own weight in a day. I have raised enough caterpillars to know that if one ate three times its own weight in a day it would have performed a skin-stretching feat. Long after writing this, but before the manuscript left my hands, I found that the origin of this statement lies in a table compiled by Trouvelot, in which he estimates that a Polyphemus caterpillar ten days old weighs one half grain, or ten times its original weight; at twenty days three grains, or sixty times its first weight; and so on until at fifty-six days it weighs two hundred and seven grains, or four thousand one hundred and forty times its first weight. To this he adds one half ounce of water and concludes: "So the food taken by a single silkworm in fifty-six days equals in weight eighty-six thousand times the primitive weight of the worm." This is a far cry from eating eighty-six thousand times its own weight in a day and upholds in part my contention in the first chapter, that people attempting to write upon these subjects "are not always rightly informed."

When the feeding period is finished in freedom, the caterpillar, if hairless, must be ready to evolve from its interior, the principal part of the winter quarters characteristic of its species while changing to the moth form, and in the case of non-feeders, sustenance for the lifetime of the moth also. Similar to the moth, the caterpillar is made up of three parts, head, thorax, and abdomen, with the organs and appendages of each. Immediately after moulting the head appears very large, and seems much too heavy for the size of the body. At the end of a feeding period and just previous to another moult the body has grown until the head is almost lost from sight, and it now seems small and insignificant; so that the appearance of a caterpillar depends on whether you examine it before or after moulting.

The head is made up of rings or segments, the same as the body, but they are so closely set that it seems to be a flat, round, or pointed formation with discernible rings on the face before casting time. The eyes are of so simple form that they are supposed only to distinguish light from darkness. The complicated mouth is at the lower part of the head. It carries a heavy pair of cutters with which the caterpillar bites off large pieces of leaf, a first pair of grinders with which it macerates the food, and a second pair that join in forming the under lip. There is also the tube that connects with the silk glands and ends in the spinneret. Through this tube a fluid is forced that by movements of the head the caterpillar attaches where it will and draws into fine threads that at once harden in silk. This organism is sufficiently developed for use in a newly emerged caterpillar, for it can spin threads by which to drop from leaf to leaf or to guide it back to a starting point.

The thorax is covered by the first three rings behind the head, and on it are six legs, two on each segment. The remainder of the caterpillar is abdominal and carries small pro-legs with which to help it cling to twigs and leaves, and the heavy anal props that support the vent. By using these and several of the pro-legs immediately before them, the caterpillar can cling and erect the front part of the body so that it can strike from side to side when disturbed. In the case of caterpillars that have a horn, as Celeus, or sets of them as Regalis, in this attitude they really appear quite formidable, and often I have seen them drive away small birds, while many people flee shrieking.

There are little tubes that carry air to the trachea, as caterpillars have no lungs and can live with a very small amount of air.

The skin may be rough, granulated, or soft and fine as silk, and in almost every instance of exquisite colour: bluish green, greenish blue, wonderful yellows and from pale to deep wine red, many species having oblique touches of contrasting colours on the abdominal rings. Others are marked with small projections of bright colours from which tufts of hair or bristles may grow. In some, as Io, these bristles are charged with an irritating acid that will sting for an hour after coming in contact with the skin, but does no permanent injury. On a few there are what seem to be small pockets of acid that can be ejected with a jerk, and on some a sort of filament that is supposed to distil a disagreeable odour. As the caterpillar only uses these when disturbed, it is safe to presume that they are placed for defence, but as in the case of moths I doubt their efficacy.

Some lepidopterists have thought the sex of a moth could be regulated by the amount of food given the caterpillar; but with my numerous other doubts I include this. It is all of a piece with any attempt at sex regulation. I regard it as morally certain that sex goes back to the ovary and that the egg produced yields a male or female caterpillar in the beginning. I am becoming convinced that caterpillars recognize sex in each other, basing the theory on the facts that in half a dozen instances I have found cocoons, spun only a few inches apart. One pair brought to me as interwoven. Two of these are shown in the following chapter. In all cases a male and female emerged within a few minutes of each other and mated as soon as possible. If a single pair of these cocoons ever had produced two of a kind, it would give rise to doubts. When all of them proved to be male and female that paired, it seems to me to furnish conclusive evidence that the caterpillars knew what they were doing, and spun in the same place for the purpose of appearing together.

At maturity, usually near five weeks, the full-fed caterpillar rests a day, empties the intestines, and races around searching for a suitable place to locate winter quarters. With burrowing caterpillars that winter in pupa cases, soft earth or rotting wood is found and entered by working their way with the heads and closing it with the hind parts. At the desired depth they push in all directions with such force that a hollow larger, but shaped as a hen's egg, is worked out; usually this is six or more inches below the surface. So compactly is the earth forced back, that fall rains, winter's alternate freezing and thawing, always a mellowing process, and spring downpours do not break up the big ball, often larger than a quart bowl, that surrounds the case of the pupa.

It has been thought by some and recorded, that this ball is held in place by spinning or an acid ejected by the caterpillar. I never have heard of any one else who has had my luck in lifting these earth balls intact, opening, and photographing them and their contents. I have examined them repeatedly and carefully. I can find not the slightest trace of spinning or adhesion other than by force.

With one of these balls lifted and divided, we decided what happened underground by detaining a caterpillar on the surface and forcing it to transform before us, for this change is not optional. When the time comes the pupa must evolve. So the caterpillar lies on the earth, gradually growing shorter, the skin appearing dry and the horns drooping. There never is a trace of spinning or acid ejected in the sand buckets. When the change is completed there begins a violent twisting and squirming. The caterpillar skin opens in a straight line just behind the head on the back, and by working with the pointed abdomen the pupa case emerges. The cast skin rapidly darkens, and as I never have found a trace of it in an opened earth ball in the spring, I suppose it disintegrates rapidly, or what is more possible, is eaten by small borers that swarm through the top six inches of the earth's crust.

The pupa is thickly coated with a sticky substance that seems to serve the double purpose of facilitating its exit from the caterpillar skin and to dry over it in a glossy waterproof coating. At first the pupa is brownish green and flattened, but as it dries it rapidly darkens in colour and assumes the shape of a perfect specimen. Concerning this stage of the evolution of a moth the doctors disagree.

The emergence I have watched repeatedly, studied photographically, and recorded in the tabulated records from which I wrote the following life histories. At time to appear I believe the pupa bores its way with the sharp point of the abdomen; at least I have seen Celeus, and Carolina, Regalis and Imperialis coming through the surface, abdomen tip first. Once free, they press with the feet against the wing shields, burst them away and leave the case at the thorax. Each moth I ever have seen emerge has been wet and the empty case damp inside. I have poured three large drops of pinkish liquid the consistency of thin cream from the abdominal rings of a Regalis case. Undoubtedly this liquid is ejected by the moth to enable it to break loose from and leave the case with its delicate down intact. The furry scales of its covering are so loosely set that any violent struggle with dry down would disfigure the moth.

Among Cecropia and its Attacine cousins, also Luna, Polyphemus, and all other spinners the process is practically the same, save that it is much more elaborate; most of all with Cecropia, that spins the largest cocoon I ever have seen, and it varies its work more than any of the others. Lengthwise of a slender twig it spins a long, slim cocoon;

on a board or wall, roomier and wider at the bottom, and inside hollow trees, and under bridges, big baggy quarters of exquisite reddish tan colours that do not fade as do those exposed to the weather. The typical cocoon of the species is that spun on a fence or outbuilding, not the slender work on the alders or the elaborate quarters of the bridge. On a board the process is to cover the space required with a fine spinning that glues firmly to the wood. Then the worker takes a firm grip with the anal props and lateral feet and begins drawing out long threads that start at the top, reach down one side, across the bottom and back to the top again, where each thread is cut and another begun. As long as the caterpillar can be seen through its work, it remains in the same position and throws the head back and around to carry the threads. I never thought of counting these movements while watching a working spinner, but some one who has, estimates that Polyphemus, that spins a cocoon not one fourth the size of Cecropia, moves the head a quarter of a million times in guiding the silk thread. When a thin webbing is spun and securely attached all around the edges it is pushed out in the middle and gummed all over the inside with a liquid glue that oozes through, coalesces and hardens in a waterproof covering. Then a big nest of crinkly silk threads averaging from three to four inches in length are spun, running from the top down one side, up the other, and the cut ends drawn closely together. One writer states that this silk has no commercial value; while Packard thinks it has. I attach greater weight to his opinion. Next comes the inner case. For this the caterpillar loosens its hold and completely surrounds itself with a small case of compact work. This in turn is saturated with the glue and forms in a thick, tough case, rough on the outside, the top not so solidly spun as the other walls; inside dark brown and worn so smooth it seems as if oiled, from the turning of the caterpillar. In this little chamber close the length and circumference of an average sized woman's two top joints of the first finger, the caterpillar transforms to the pupa stage, crowding its cast skin in a wad at the bottom.

At time for emergence the moth bursts the pupa case, which is extremely thin and papery compared with the cases of burrowing species. We know by the wet moth that liquid is ejected, although we cannot see the wet spot on the top of the inner case of Cecropia as we can with Polyphemus, that does not spin the loose outer case and silk nest. From here on the moths emerge according to species. Some work with their mouths and fore feet. Some have rough projections on the top of the head, and others little sawlike arrangements at the bases of the wings. In whatever manner they free themselves, all of them are wet when they leave their quarters. Sometimes the gathered silk ends comb sufficient down from an emerging Cecropia to leave a terra cotta rim around the opening from which it came; but I never saw one lose enough at this time to disfigure it. On very rare occasions a deformed moth appears. I had a Cecropia with one wing no larger than my thumb nail, and it never developed. This is caused by the moth sustaining an injury to the wing in emergence. If the membrane is slightly punctured the liquid forced into the wing for its development escapes and there is no enlargement.

Also, in rare instances, a moth is unable to escape at all and is lost if it is not assisted; but this is precarious business and should not be attempted unless you are positive the moth will die if you do not interfere. The struggle it takes to emerge is a part of the life process of the moth and quickens its circulation and develops its strength for the affairs of life afterward. If the feet have a steady pull to drag forth the body, they will be strong enough to bear its weight while the wings dry and develop.

All lepidopterists mention the wet condition of the moths when they emerge. Some explain that an acid is ejected to soften the pupa case so that the moth can cut its way out; others go a step farther and state that the acid is from the mouth. I am extremely curious about this. I want to know just what this acid is and where it comes from. I know of no part of the thorax provided with a receptacle for the amount of liquid used to flood a case, dampen a moth, and leave several drops in the shell.

As soon as a moth can find a suitable place to cling after it is out, it hangs by the feet and dries the wings and down. Long before it is dry if you try to move a moth or cause disturbance, it will eject several copious jets of a spray from the abdomen that appears, smells and tastes precisely like the liquid found in the abandoned case. If protected from the lightest touch it will do the same. It appeals to me that this liquid is abdominal, partly thrown off to assist the moth in emergence; something very like that bath of birth which accompanies and facilitates human entrance into the world. It helps the struggling moth in separating from the case, wets the down so that it will pass the small opening, reduces the large abdomen so that it will escape the exit, and softens the case and silk where the moth is working. With either male or female the increase in size is so rapid that neither could be returned to their cases five minutes after they have left them.

It is generally supposed that the spray thrown by a developing moth is for the purpose of attracting others of its kind. I have my doubts. With moths that have been sheltered and not even touched by a breath of wind, this spray is thrown very frequently before the moth is entirely dry, long before it is able to fly and before the ovipositor is thrust out. According to my sense of smell there is very little odour to the spray and what there is would be dissipated hours before night and time for the moths to fly and seek mates. I do not think that the spray thrown so soon after escape from cocoon or case is to attract the sexes, any farther than that much of it in one place on something that it would saturate might leave a general 'mothy' odour. Some lepidopterists think this spray a means of defence; if this is true I fail to see why it should be thrown when there is nothing disturbing the moth.

Many of the spinning moths use leaves for their outer foundation. Some appear as if snugly rolled in a leaf and hanging from a twig, but examination will prove that the stem is silk covered to hold the case when the leaf loosens. This is the rule with all Promethea cocoons I ever have seen. Polyphemus selects a cluster of leaves very frequently thorn, and weaves its cocoon against three, drawing them together and spinning a support the length of the stems, so that when the leaf is ready to fall the cocoon is safely anchored. When the winter winds have beaten the edges from the leaves, the cocoon appears as if it were brown, having three ribs with veins running from them, and of triangular shape. Angulifera spins against the leaves but provides no support and so drops to the ground. Luna spins a comparatively thin white case, among the leaves under the shelter of logs and stumps. Io spins so slightly in confinement that the pupa case and cast skin show through. I never have found a pupa out of doors, but this is a ground caterpillar.

Sometimes the caterpillar has been stung and bad an egg placed in its skin by a parasite, before pupation. In such case the pupa is destroyed by the developing fly. Throughout one winter I was puzzled by the light weight of what appeared to be a good Polyphemus cocoon, and at time for emergence amazed by the tearing and scratching inside the cocoon, until what I think was an Ophion fly appeared. It was honey yellow, had antennae long as its extremely long body, the abdomen of which was curved and the segments set together so as to appear notched. The wings were transparent and the insect it seems is especially designed to attack Polyphemus caterpillars and help check a progress that otherwise might become devastating.

Among the moths that do not feed, the year of their evolution is divided into about seven days for the life of the moth, from fifteen to thirty for the eggs, from five to six weeks for the caterpillar and the remainder of the time in the pupa stage. The rule differs with feeding moths only in that after mating and egg placing they take food and live several months, often until quite heavy frosts have fallen.

One can admire to fullest extent the complicated organism, wondrous colouring, and miraculous life processes in the evolution of a moth, but that is all. Their faces express nothing; their attitudes tell no story. There is the marvellous instinct through which the males locate the opposite sex of their species; but one cannot see instinct in the face of any creature; it must develop in acts. There is no part of their lives that makes such pictures of mother-love as birds and animals afford. The male finds a mate and disappears. The female places her eggs and goes out before her caterpillars break their shells. The caterpillar transforms to the moth without its consent, the matter in one upbuilding the other. The entire process is utterly devoid of sentiment, attachment or volition on the part of the creatures involved. They work out a law as inevitable as that which swings suns, moons, and planets in their courses. They are the most fragile and beautiful result of natural law with which I am acquainted.

CHAPTER III.
THE ROBIN MOTH: CECROPIA

When only a little child, wandering alone among the fruits and flowers of our country garden, on a dead peach limb beside the fence I found it—my first Cecropia. I was the friend of every bird, flower, and butterfly. I carried crumbs to the warblers in the sweetbrier; was lifted for surreptitious peeps at the hummingbird nesting in the honeysuckle; sat within a few feet of the robin in the catalpa; bugged the currant bushes for the phoebe that had built for years under the roof of the corn bin; and fed young blackbirds in the hemlock with worms gathered from the cabbages. I knew how to insinuate myself into the private life of each bird that homed on our farm, and they were many, for we valiantly battled for their protection with every kind of intruder. There were wrens in the knot holes, chippies in the fences, thrushes in the brush heaps, bluebirds in the hollow apple trees, cardinals in the bushes, tanagers in the saplings, fly-catchers in the trees, larks in the wheat, bobolinks in the clover, killdeers beside the creeks, swallows in the chimneys, and martins under the barn eaves. My love encompassed all feathered and furred creatures.

Every day visits were paid flowers I cared for most. I had been taught not to break the garden blooms, and if a very few of the wild ones were taken, I gathered them carefully, and explained to the plants that I wanted them for my mother because she was so ill she could not come to them any more, and only a few touching her lips or lying on her pillow helped her to rest, and made vivid the fields and woods when the pain was severe.

My love for the butterflies took on the form of adoration. There was not a delicate, gaudy, winged creature of day that did not make so strong an appeal to my heart as to be almost painful. It seemed to me that the most exquisite thoughts of God for our pleasure were materialized in their beauty. My soul always craved colour, and more brilliancy could be found on one butterfly wing than on many flower faces. I liked to slip along the bloom-bordered walks of that garden and stand spell-bound, watching a black velvet butterfly, which trailed wings painted in white, red, and green, as it clambered over a clump of sweet-williams, and indeed, the flowers appeared plain compared with it! Butterflies have changed their habits since then. They fly so high! They are all among the treetops now. They used to flit around the cinnamon pinks, larkspur, ragged-robins and tiger lilies, within easy reach of little fingers, every day. I called them 'flying flowers,' and it was a pretty conceit, for they really were more delicate in texture and brighter in colouring than the garden blooms.

Having been taught that God created the heavens, earth and all things therein, I understood it to mean a literal creation of each separate thing and creature, as when my father cut down a tree and hewed it into a beam. I would spend hours sitting so immovably among the flowers of our garden that the butterflies would mistake me for a plant and alight on my head and hands, while I strove to conceive the greatness of a Being who could devise and colour all those different butterfly wings. I would try to decide whether He created the birds, flowers, or butterflies first; ultimately coming to the conclusion that He put His most exquisite material into the butterflies, and then did the best He could with what remained, on the birds and flowers.

In my home there was a cellar window on the south, covered with wire screening, that was my individual property. Father placed a box beneath it so that I could reach the sill easily, and there were very few butterflies or insects common to eastern North America a specimen of which had not spent some days on that screen, feasted on leaves and flowers, drunk from saucers of sweetened water, been admired and studied in minutest detail, and then set free to enjoy life as before. With Whitman, "I never was possessed with a mania for killing things." I had no idea of what families they were, and I supplied my own names. The Monarch was the Brown Velvet; the Viceroy was his Cousin; the Argynnis was the Silver Spotted; and the Papilio Ajax was the Ribbon butterfly, in my category. There was some thought of naming Ajax, Dolly Varden; but on close inspection it seemed most to resemble the gayly striped ribbons my sisters wore.

I was far afield as to names, but in later years with only a glance at any specimen I could say, "Oh, yes! I always have known that. It has buff-coloured legs, clubbed antennae with buff tips, wings of purplish brown velvet with escalloped margins, a deep band of buff lightly traced with black bordering them, and a pronounced point close the apex of the front pair. When it came to books, all they had to teach me were the names. I had captured and studied butterflies, big, little, and with every conceivable variety of marking, until it was seldom one was found whose least peculiarity was not familiar to me as my own face; but what could this be?

It clung to the rough bark, slowly opening and closing large wings of grey velvet down, margined with bands made of shades of grey, tan, and black; banded with a broad stripe of red terra cotta colour with an inside margin of white, widest on the back pair. Both pairs of wings were decorated with half-moons of white, outlined in black and strongly flushed with terra cotta; the front pair near the outer margin had oval markings of blue-black, shaded with grey, outlined with half circles of white, and secondary circles of black. When the wings were raised I could see a face of terra cotta, with small eyes, a broad band of white across the forehead, and an abdomen of terra cotta banded with snowy white above, and spotted with white beneath. Its legs were hairy, and the antennae antlered like small branching ferns. Of course I thought it was a butterfly, and for a time was too filled with wonder to move. Then creeping close, the next time the wings were raised above its body, with the nerveless touch of a robust child I captured it.

I was ten miles from home, but I had spent all my life until the last year on that farm, and I knew and loved every foot of it. To leave it for a city home and the confinement of school almost had broken my heart, but it really was time for me to be having some formal education. It had been the greatest possible treat to be allowed to return to the country for a week, but now my one idea was to go home with my treasure. None of my people had seen a sight like that. If they had, they would have told me.

Borrowing a two-gallon stone jar from the tenant's wife, I searched the garden for flowers sufficiently rare for lining. Nothing so pleased me as some gorgeous deep red peony blooms. Never having been allowed to break the flowers when that was my mother's home, I did not think of doing it because she was not there to know. I knelt and gathered all the fallen petals that were fresh, and then spreading my apron on the ground, jarred the plant, not harder than a light wind might, and all that fell in this manner it seemed right to take. The selection was very pleasing, for the yellow glaze of the jar, the rich red of the petals, and the grey velvet of my prize made a picture over which I stood trembling in delight. The moth was promptly christened the Half-luna, because my father had taught me that luna was the moon, and the half moons on the wings were its most prominent markings.

The tenant's wife wanted me to put it in a pasteboard box, but I stubbornly insisted on having the jar, why, I do not know, but I suppose it was because my father's word was gospel to me, and he had said that the best place to keep my specimens was the cellar window, and I must have thought the jar the nearest equivalent to the cellar. The Half-luna did not mind in the least, but went on lazily opening and closing its wings, yet making no attempt to fly. If I had known what it was, or anything of its condition, I would have understood that it had emerged from the cocoon that morning, and never had flown, but was establishing circulation preparatory to taking wing. Being only a small, very ignorant girl, the greatest thing I knew for sure was what I loved.

Tying my sunbonnet over the top of the jar, I stationed myself on the horse block at the front gate. Every passing team was hailed with lifted hand, just as I had seen my father do, and in as perfect an imitation of his voice as a scared little girl making her first venture alone in the big world could muster, I asked, "Which way, Friend?"

For several long, hot hours people went to every point of the compass, but at last a bony young farmer, with a fat wife, and a fatter baby, in a big wagon, were going to my city, and they said I might ride. With quaking heart I handed up my jar, and climbed in, covering all those ten miles in the June sunshine, on a board laid across e wagon bed, tightly clasping the two-gallon jar in my aching arms. The farmer's wife was quite concerned about me. She asked if I had butter, and I said, "Yes, the kind that flies."

I slipped the bonnet enough to let them peep. She did not seem to think much of it, but the farmer laughed until his tanned face was red as an Indian's. His wife insisted on me putting down the jar, and offered to set her foot on it so that it would not 'jounce' much, but I did not propose to risk it 'jouncing' at all, and clung to it persistently. Then she offered to tie her apron over the top of the jar if I would put my bonnet on my head, but I was afraid to attempt the exchange for fear my butterfly would try to escape, and I might crush it, a thing I almost never had allowed to happen.

The farmer's wife stuck her elbow into his ribs, and said, "How's that for the queerest spec'men ye ever see?" The farmer answered, "I never saw nothin' like it before." Then she said, "Aw pshaw! I didn't mean in the jar!" Then they both laughed. I thought they were amused at me, but I had no intention of risking an injury to my Half-luna, for there had been one black day on which I had such a terrible experience that it entailed a lifetime of caution.

I had captured what I afterward learned was an Asterias, that seemed slightly different from any previous specimen, and a yellow swallow-tail, my first Papilio Turnus. The yellow one was the largest, most beautiful butterfly I ever had seen. I was carrying them, one between each thumb and forefinger, and running with all possible speed to reach the screen before my touch could soil the down on their exquisite wings. I stumbled, and fell, so suddenly, there was no time to release them. The black one sailed away with a ragged wing, and the yellow was crushed into a shapeless mass in my hand. I was accustomed to falling off fences, from trees, and into the creek, and because my mother was an invalid I had learned to doctor my own bruises and uncomplainingly go my way. My reputation was that of a very brave little girl; but when I opened my hand and saw that broken butterfly, and my down-painted fingers, I was never more afraid in my life. I screamed aloud in panic, and ran for my mother with all my might. Heartbroken, I could not control my voice to explain as I threw myself on her couch, and before I knew what they were doing, I was surrounded by sisters and the cook with hot water, bandages and camphor.

My mother clasped me in her arms, and rocked me on her breast. "There, there, my poor child," she said, "I know it hurts dreadfully!" And to the cook she commanded, "Pour on camphor quickly! She is half killed, or she never would come to me like this." I found my voice. "Camphor won't do any good," I wailed. "It was the most beautiful butterfly, and I've broken it all to pieces. It must have taken God hours studying how to make it different from all the others, and I know He never will forgive me!" I began sobbing worse than ever. The cook on her knees before me sat on her heels suddenly. "Great Heavens! She's screechin' about breakin' a butterfly, and not her poor fut, at all!" Then I looked down and discovered that I had stubbed my toe in falling, and had left a bloody trail behind me. "Of course I am!" I sobbed indignantly. "Couldn't I wash off a little blood in the creek, and tie up my toe with a dock leaf and some grass? I've killed the most beautiful butterfly, and I know I won't be forgiven!"

I opened my tightly clenched hand and showed it to prove my words. The sight was so terrible to me that I jerked my foot from the cook, and thrust my hand into the water, screaming, "Wash it! Wash it! Wash the velvet from my hand! Oh! make it white again!" Before the cook bathed and bandaged my foot, she washed and dried my hand; and my mother whispered, "God knows you never meant to do it, and He is sorry as mother is." So my mother and the cook comforted me. The remainder scattered suddenly. It was years before I knew why, and I was a Shakespearean student before I caught the point to their frequently calling me 'Little Lady

Macbeth!' After such an experience, it was not probable that I would risk crushing a butterfly to tie a bonnet on my head. It probably would be down my back half the time anyway. It usually was. As we neared the city I heard the farmer's wife tell him that he must take me to my home. He said he would not do any such a thing, but she said he must. She explained that she knew me, and it would not be decent to put me down where they were going, and leave me to walk home and carry that heavy jar. So the farmer took me to our gate. I thanked him as politely as I knew how, and kissed his wife and the fat baby in payment for their kindness, for I was very grateful. I was so tired I scarcely could set down the jar and straighten my cramped arms when I had the opportunity. I had expected my family to be delighted over my treasure, but they exhibited an astonishing indifference, and were far more concerned over the state of my blistered face. I would not hear of putting my Half-luna on the basement screen as they suggested, but enthroned it in state on the best lace curtains at a parlour window, covered the sill with leaves and flowers, and went to bed happy. The following morning my sisters said a curtain was ruined, and when they removed it to attempt restoration, the general consensus of opinion seemed to be that something was a nuisance, I could not tell whether it was I, or the Half-luna. On coming to the parlour a little later, ladened with leaves and flowers, my treasure was gone. The cook was sure it had flown from the door over some

58

one's head, and she said very tersely that it was a burning shame, and if such carelessness as that ever occurred again she would quit her job. Such is the confidence of a child that I accepted my loss as an inevitable accident, and tried to be brave to comfort her, although my heart was almost broken. Of course they freed my moth. They never would have dared but that the little mother's couch stood all day empty now, and her chair unused beside it. My disappointment was so deep and far-reaching it made me ill then they scolded me, and said I had half killed myself carrying that heavy jar in the hot sunshine, although the pain from which I suffered was neither in my arms nor sunburned face.

So I lost my first Cecropia, and from that day until a woman grown and much of this material secured, in all my field work among the birds, flowers, and animals, I never had seen another. They had taunted me in museums, and been my envy in private collections, but find one, I could not. When in my field work among the birds, so many moths of other families almost had thrust themselves upon me that I began a collection of reproductions of them, I found little difficulty in securing almost anything else. I could picture Sphinx Moths in any position I chose, and Lunas seemed eager to pose for me. A friend carried to me a beautiful tan-coloured Polyphemus with transparent moons like isinglass set in its wings of softest velvet down, and as for butterflies, it was not necessary to go afield for them; they came to me.

I could pick a Papilio Ajax, that some of my friends were years in securing, from the pinks in my garden. A pair of Antiopas spent a night, and waited to be pictured in the morning, among the leaves of my passion vine. Painted Beauties swayed along my flowered walks, and in September a Viceroy reigned in state on every chrysanthemum, and a Monarch was enthroned on every sunbeam. No luck was too good for me, no butterfly or moth too rare, except forever and always the coveted Cecropia, and by this time I had learned to my disgust that it was one of the commonest of all.

Then one summer, late in June, a small boy, having an earnest, eager little face, came to me tugging a large box. He said he had something for me. He said "they called it a butterfly, but he was sure it never was." He was eminently correct. He had a splendid big Cecropia. I was delighted. Of course to have found one myself would have filled my cup to overflowing, but to secure a perfect, living specimen was good enough. For the first time my childish loss seemed in a measure compensated. Then, I only could study a moth to my satisfaction and set it free; now, I could make reproductions so perfect that every antler of its antennae could be counted with the naked eye, and copy its colours accurately, before giving back its liberty.

I asked him whether he wanted money or a picture of it, and as I expected, he said 'money,' so he was paid. An hour later he came back and said he wanted the picture. On being questioned as to his change of heart, he said "mamma told him to say he wanted the picture, and she would give him the money." My sympathy was with her. I wanted the studies I intended to make of that Cecropia myself, and I wanted them very badly.

I opened the box to examine the moth, and found it so numb with the cold over night, and so worn and helpless, that it could not cling to a leaf or twig. I tried repeatedly, and fearing that it had been subjected to rough treatment, and soon would be lifeless, for these moths live only a short time, I hastily set up a camera focusing on a branch. Then I tried posing my specimen. Until the third time it fell, but the fourth it clung, and crept down a twig, settling at last in a position that far, surpassed any posing that I could do. I was very pleased, and yet it made a complication. It had gone so far that it might be off the plate and from focus. It seemed so stupid and helpless that I decided to risk a peep at the glass, and hastily removing the plate and changing the shutter, a slight but most essential alteration was made, everything replaced, and the bulb caught up. There was only a breath of sound as I turned, and then I stood horrified, for my Cecropia was sailing over a large elm tree in a corner of the orchard, and for a block my gaze followed it skyward, flying like a bird before it vanished in the distance, so quickly had it recovered in fresh air and sunshine.

I have undertaken to describe some very difficult things, but I would not attempt to portray my feelings, and three days later there was no change. It was in the height of my season of field work, and I had several extremely interesting series of bird studies on hand, and many miscellaneous subjects. In those days some pictures were secured that I then thought, and yet feel, will live, but nothing mattered to me. There was a standing joke among my friends that I never would be satisfied with my field work until I had made a study of a 'Ha-ha bird,' but I doubt if even that specimen would have lifted the gloom of those days. Everything was a drag, and frequently I would think over it all in detail, and roundly bless myself for taking a prize so rare, to me at least, into the open.

The third day stands lurid in my memory. It was the hottest, most difficult day of all my years of experience afield. The temperature ranged from 104 to 108 in the village, and in quarries open to the east, flat fields, and steaming swamps it certainly could have been no cooler. With set cameras I was working for a shot at a hawk that was feeding on all the young birds and rabbits in the vicinity of its nest. I also wanted a number of studies to fill a commission that was pressing me. Subjects for several pictures had been found, and exposures made on them when the weather was so hot that the rubber slide of a plate holder would curl like a horseshoe if not laid on a case, and held flat by a camera while I worked.

Perspiration dried, and the landscape took on a sombre black velvet hue, with a liberal sprinkling of gold stars. I sank into a stupor going home, and an old farmer aroused me, and disentangled my horse from a thicket of wild briers into which it had strayed. He said most emphatically that if I did not know enough to remain indoors weather like that, my friends should appoint me a 'guardeen.'

I reached the village more worn in body and spirit than I ever had been. I felt that I could not endure another degree of heat on the back of my head, and I was much discouraged concerning my work. Why not drop it all, and go where there were cool forests and breezes sighing? Perhaps my studies were not half so good as I thought! Perhaps people would not care for them! For that matter, perhaps the editors and publishers never would give the public an opportunity to see my work at all!

I dragged a heavy load up the steps and swung it to the veranda, and there stood almost paralysed. On the top step, where I could not reach the Cabin door without seeing it, newly emerged, and slowly exercising a pair of big wings, with every gaudy marking fresh with new life, was the finest Cecropia I ever had seen anywhere. Recovering myself with a start, I had it under my net that had waited twenty years to cover it! Inside the door I dropped the net, and the moth crept on my fingers. What luck! What extra golden luck! I almost felt that God had been sorry for me, and sent it there to encourage me to keep on picturing the beauties and wonders of His creations for people who could not go afield to see for themselves, and to teach those who could to protect helpless, harmless things for their use and beauty.

I walked down the hall, and vaguely scanned the solid rows of books and specimens lining the library walls. I scarcely realized the thought that was in my mind, but what I was looking for was not there. The dining-room then, with panelled walls and curtains of tapestry? It was not there! Straight to the white and gold music room I went. Then a realizing sense came to me. It was BRUSSELS LACE for which I was searching! On the most delicate, snowiest place possible, on the finest curtain there, I placed my Cecropia, and then stepped back and gazed at it with a sort of "Touch it over my dead body" sentiment in my heart. An effort was required to arouse myself, to realize that I was not dreaming. To search the fields and woods for twenty years, and then find the specimen I had sought awaiting me at my own door! Well might it have been a dream, but that the Cecropia, clinging to the meshes of the lace, slowly opening and closing its wings to strengthen them for flight, could be nothing but a delightful reality.

A few days later, in the valley of the Wood Robin, while searching for its nest I found a large cocoon. It was above my head, but afterward I secured it by means of a ladder, and carried it home. Shortly there emerged a yet larger Cecropia, and luck seemed with me. I could find them everywhere through June, the time of their emergence, later their eggs, and the tiny caterpillars that hatched from them. During the summer I found these caterpillars, in different stages of growth, until fall, when after their last moult and casting of skin, they reached the final period of feeding; some were over four inches in length, a beautiful shade of greenish blue, with red and yellow warty projections—tubercles, according to scientific works.

It is easy to find the cocoons these caterpillars spin, because they are the largest woven by any moth, and placed in such a variety of accessible spots. They can be found in orchards, high on branches, and on water sprouts at the base of trees. Frequently they are spun on swamp willows, box-elder, maple, or wild cherry. Mr. Black once found for me the largest cocoon I ever have seen; a pale tan colour with silvery lights, woven against the inside of a hollow log. Perhaps the most beautiful of all, a dull red, was found under the flooring of an old bridge crossing a stream in the heart of the swamp, by a girl not unknown to fiction, who brought it to me. In a deserted orchard close the Wabash, Raymond once found a pair of empty cocoons at the foot of a big apple tree, fastened to the same twigs, and within two inches of each other.

But the most wonderful thing of all occurred when Wallace Hardison, a faithful friend to my work, sawed a board from the roof of his chicken house and carried to me twin Cecropia cocoons, spun so closely together they were touching, and slightly interwoven. By the closest examination I could discover slight difference between them. The one on the right was a trifle fuller in the body, wider at the top, a shade lighter in colour, and the inner case seemed heavier.

All winter those cocoons occupied the place of state in my collection. Every few days I tried them to see if they gave the solid thump indicating healthy pupae, and listened to learn if they were moving. By May they were under constant surveillance. On the fourteenth I was called from home a few hours to attend the funeral of a friend. I think nothing short of a funeral would have taken me, for the moth from a single cocoon had emerged on the eleventh. I hurried home near noon, only to find that I was late, for one was out, and the top of the other cocoon heaving with the movements of the second.

The moth that had escaped was a male. It clung to the side of the board, wings limp, its abdomen damp. The opening from which it came was so covered with terra cotta coloured down that I thought at first it must have disfigured itself; but full development proved it could spare that much and yet appear all right.

In the fall I had driven a nail through one corner of the board, and tacked it against the south side of the Cabin, where I made reproductions of the cocoons. The nail had been left, and now it suggested the same place. A light stroke on the head of the nail, covered with cloth to prevent jarring, fastened the board on a log. Never in all my life did I hurry as on that day, and I called my entire family into service. The Deacon stood at one elbow, Molly-Cotton at the other, and the gardener in the rear. There was not a second to be lost, and no time for an unnecessary movement; for in the heat and bright sunshine those moths would emerge and develop with amazing rapidity.

Molly-Cotton held an umbrella over them to prevent this as much as possible; the Deacon handed plate holders, and Brenner ran errands. Working as fast as I could make my fingers fly in setting up the camera, and getting a focus, the second moth's head was out, its front feet struggling to pull up the body; and its antennae beginning to lift, when I was ready for the first snap at half-past eleven.

By the time I inserted the slide, turned the plate holder and removed another slide, the first moth to appear had climbed up the board a few steps, and the second was halfway out. Its antennae were nearly horizontal now, and from its position I decided that the wings as they lay in the pupa case were folded neither to the back nor to the front, but pressed against the body in a lengthwise crumpled mass, the heavy front rib, or costa, on top.

Again I changed plates with all speed. By the time I was ready for the third snap the male had reached the top of the board, its wings opened for the first time, and began a queer trembling motion. The second one had emerged and was running into the first, so I held my finger in the line of its advance, and when it climbed on I lowered it to the edge to the board beside the cocoons. It immediately clung to the wood. The big pursy abdomen and smaller antennae, that now turned forward in position, proved this a female. The exposure was made not ten seconds after she cleared the case, and with her back to the lens, so the position and condition of the wings and antennae on emergence can be seen clearly.

Quickly as possible I changed the plates again; the time that elapsed could not have been over half a minute. The male was trying to creep up the wall, and the increase in the length and expansion of the female's wings could be seen. The colours on both were exquisite, but they grew a trifle less brilliant as the moths became dry.

Again I turned to the business of plate changing. The heat was intense, and perspiration was streaming from my face. I called to Molly-Cotton to shield the moths while I made the change. "Drat the moths!" cried the Deacon. "Shade your mother!" Being an obedient girl, she shifted the umbrella, and by the time I was ready for business, the male was on the logs and travelling up the side of the Cabin. The female was climbing toward the logs also, so that a side view showed her wings already beginning to lift above her back.

I had only five snapshot plates in my holders, so I was compelled to stop. It was as well, for surely the record was complete, and I was almost prostrate with excitement and heat. Several days later I opened each of the cocoons and made interior studies. The one on the right was split down the left side and turned back to show the bed of spun silk of exquisite colour that covers the inner case. Some say this silk has no commercial value, as it is cut in lengths reaching from the top around the inner case and back to the top again; others think it can be used. The one on the left was opened down the front of the outer case, the silk parted and the heavy inner case cut from top to bottom to show the smooth interior wall, the thin pupa case burst by the exit of the moth, and the cast caterpillar skin crowded at the bottom.

The pair mated that same night, and the female began laying eggs by noon the following day. She dotted them in lines over the inside of her box, and on leaves placed in it, and at times piled them in a heap instead of placing them as do these moths in freedom. Having taken a picture of a full-grown caterpillar of this moth brought to me by Mr. Andrew Idlewine, I now had a complete Cecropia history; eggs, full-grown caterpillars, twin cocoons, and the story of the emergence of the moths that wintered in them. I do not suppose Mr. Hardison thought he was doing anything unusual when he brought me those cocoons, yet by bringing them, he made it possible for me to secure this series of twin Cecropia moths, male and female, a thing never before recorded by lepidopterist or photographer so far as I can learn.

The Cecropia is a moth whose acquaintance nature-loving city people can cultivate. In December of 1906, on a tree, maple I think, near No. 2230 North Delaware Street, Indianapolis, I found four cocoons of this moth, and on the next tree, save one, another. Then I began watching, and in the coming days I counted them by the hundred through the city. Several bushels of these cocoons could have been clipped in Indianapolis alone, and there is no reason why any other city that has maple, elm, catalpa, and other shade trees would not have as many; so that any one who would like can find them easily.

Cecropia cocoons bewilder a beginner by their difference in shape. You cannot determine the sex of the moth by the size of the cocoon. In the case of the twins, the cocoon of the female was the larger; but I have known male and female alike to emerge from large or small. You are fairly sure of selecting a pair if you depend upon weight. The females are heavier than the males, because they emerge with quantities of eggs ready to deposit as soon as they have mated. If any one wants to winter a pair of moths, they are reasonably sure of doing so by selecting the heaviest and lightest cocoons they can find.

In the selection of cocoons, hold them to the ear, and with a quick motion reverse them end for end. If there is a dull, solid thump, the moth is alive, and will emerge all right. If this thump is lacking, and there is a rattle like a small seed shaking in a dry pod, it means that the caterpillar has gone into the cocoon with one of the tiny parasites that infest these worms, clinging to it, and the pupa has been eaten by the parasite.

In fall and late summer are the best times to find cocoons, as birds tear open many of them in winter; and when weatherbeaten they fade, and do not show the exquisite shadings of silk of those newly spun. When fresh, the colours range from almost white through lightest tans and browns to a genuine red, and there is a silvery effect that is lovely on some of the large, baggy ones, hidden under bridges. Out of doors the moths emerge in middle May or June, but they are earlier in the heat of a house. They are the largest of any species, and exquisitely coloured, the shades being strongest on the upper side of the wings. They differ greatly in size, most males having an average wing sweep of five inches, and a female that emerged in my conservatory from a cocoon that I wintered with particular care had a spread of seven inches, the widest of which I have heard; six and three quarters is a large female. The moth, on appearing, seems all head and abdomen, the wings hanging limp and wet from the shoulders. It at once creeps around until a place where it can hang with the wings down is found, and soon there begins a sort of pumping motion of the body. I imagine this is to start circulation, to exercise parts, and force blood into the wings. They begin to expand, to dry, to take on colour with amazing rapidity, and as soon as they are full size and crisp, the moth commences raising and lowering them slowly, as in flight. If a male, he emerges near ten in the forenoon, and flies at dusk in search of a mate.

As the females are very heavy with eggs, they usually remain where they are. After mating they begin almost at once to deposit their eggs, and do not take flight until they have finished. The eggs are round, having a flat top that becomes slightly depressed as they dry. They are of pearl colour, with a touch of brown, changing to greyish as the tiny caterpillars develop. Their outline can be traced through the shell on which they make their first meal when they emerge. Female Cecropas average about three hundred and fifty eggs each, that they sometimes place singly, and again string in rows, or in captivity pile in heaps. In freedom they deposit the eggs mostly on leaves, sometimes the under, sometimes the upper, sides or dot them on bark, boards or walls. The percentage of loss of eggs and the young is large, for they are nowhere numerous enough to become a pest, as they certainly would if three hundred caterpillars survived to each female moth. The young feed on apple, willow, maple, box-elder, or wild cherry leaves; and grow through a series of feeding periods and moults, during which they rest for a few days, cast the skin and intestinal lining and then feed for another period.

After the females have finished depositing their eggs, they cling to branches, vines or walls a few days, fly aimlessly at night and then pass out without ever having taken food.

Cecropia has several 'Cousins,' Promethea, Angulifera, Gloveri, and Cynthia, that vary slightly in marking and more in colour. All are smaller than Cecropia. The male of Promethea is the darkest moth of the Limberlost. The male of Angulifera is a brownish grey, the female reddish, with warm tan colours on her wing borders. She is very beautiful. The markings on the wings of both are not half-moon shaped, as Cecropia and Gloveri, but are oblong, and largest at the point next the apex of the wing.

Gloveri could not be told from Cecropiain half-tone reproduction by any save a scientist, so similar are the markings, but in colour they are vastly different, and more beautiful. The only living Gloveri I ever secured was almost done with life, and she was so badly battered I could not think of making a picture of her. The wings are a lovely red wine colour, with warm tan borders, and the crescents are white, with a line of tan and then of black. The abdomen is white striped with wine and black.

Cynthia has pale olive green shadings on both male and female. These are imported moths brought here about 1861 in the hope that they would prove valuable in silk culture. They occur mostly where the ailanthus grows.

My heart goes out to Cecropia because it is such a noble, birdlike, big fellow, and since it has decided to be rare with me no longer, all that is necessary is to pick it up, either in caterpillar, cocoon, or moth, at any season of the year, in almost any location. The Cecropia moth resembles the robin among birds; not alone because he is grey with red markings, but also he haunts the same localities. The robin is the bird of the eaves, the back door, the yard and orchard. Cecropia is the moth. My doorstep is not the only one they grace; my friends have found them in like places. Cecropia cocoons are attached to fences, chicken-coops, barns, houses, and all through the orchards of old country places, so that their emergence at bloom time adds to May and June one more beauty, and frequently I speak of them as the Robin Moth.

In connexion with Cecropia there came to me the most delightful experience of my life. One perfect night during the middle of May, all the world white with tree bloom, touched to radiance with brilliant moonlight; intoxicating with countless blending perfumes, I placed a female Cecropia on the screen of my sleeping-room door and retired. The lot on which the Cabin stands is sloping, so that, although the front foundations are low, my door is at least five feet above the ground, and opens on a circular porch, from which steps lead down between two apple trees, at that time sheeted in bloom. Past midnight I was awakened by soft touches on the screen, faint pullings at the wire. I went to the door and found the porch, orchard, and night-sky alive with Cecropias holding high carnival. I had not supposed there were so many in all this world. From every direction they came floating like birds down the moonbeams. I carefully removed the female from the door to a window close beside, and stepped on the porch. No doubt I was permeated with the odour of the moth. As I advanced to the top step, that lay even with the middle branches of the apple trees, the exquisite big creatures came swarming around me. I could feel them on my hair, my shoulders, and see them settling on my gown and outstretched hands.

Far as I could penetrate the night-sky more were coming. They settled on the bloom-laden branches, on the porch pillars, on me indiscriminately. I stepped inside the door with one on each hand and five clinging to my gown. This experience, I am sure, suggested Mrs. Comstock's moth hunting in the Limberlost. Then I went back to the veranda and revelled with the moths until dawn drove them to shelter. One magnificent specimen, birdlike above all the others, I followed across the orchard and yard to a grape arbour, where I picked him from the under side of a leaf after he had settled for the coming day. Repeatedly I counted close to a hundred, and then they would so confuse me by flight I could not be sure I was not numbering the same one twice. With eight males, some of them fine large moths, one superb, from which to choose, my female mated with an insistent, frowsy little scrub lacking two feet and having torn and ragged wings. I needed no surer proof that she had very dim vision.

CHAPTER IV.
THE YELLOW EMPEROR:
EACLES IMPERIALIS

Several years ago, Mr. A. Eisen, a German, of Coldwater, Michigan, who devotes his leisure to collecting moths, gave me as pinned specimens a pair of Eacles Imperialis, and their full life history. Any intimate friend of mine can testify that yellow is my favourite colour, with shades of lavender running into purple, second choice. When I found a yellow moth, liberally decorated with lavender, the combination was irresistible. Mr. Eisen said the mounted specimens were faded; but the living moths were beautiful beyond description. Naturally I coveted life.

I was very particular to secure the history of the caterpillars and their favourite foods. I learned from Mr. Eisen that they were all of the same shape and habit, but some of them might be green, with cream-coloured heads and feet, and black face lines, the body covered sparsely with long hairs; or they might be brown, with markings of darker brown and black with white hairs; but they would be at least three inches long when full grown, and would have a queer habit of rearing and drawing leaves to their mouths when feeding. I was told I would find them in August, on leaves of spruce, pine, cherry, birch, alder, sycamore, elm, or maple; that they pupated in the ground; and the moths were common, especially around lights in city parks, and at street crossings.

Coming from a drive one rare June evening, I found Mr. William Pettis, a shooter of oil wells, whom I frequently met while at my work, sitting on the veranda in an animated business discussion with the Deacon.

"I brought you a pair of big moths that I found this morning on some bushes beside the road," said Mr. Pettis. "I went to give Mr. Porter a peep to see if he thought you'd want them, and they both got away. He was quicker than I, and caught the larger one, but mine sailed over the top of that tree." He indicated an elm not far away.

"Did you know them?" I asked the Deacon.

"No," he answered. "You have none of the kind. They are big as birds and a beautiful yellow."

"Yellow!" No doubt I was unduly emphatic. "Yellow! Didn't you know better than to open a box with moths in it outdoors at night?"

"It was my fault," interposed Mr. Pettis. "He told me not to open the box, but I had shown them a dozen times to-day and they never moved. I didn't think about night being their time to fly. I am very sorry."

So was I. Sorry enough to have cried, but I tried my best to conceal it. Anyway, it might be Io, and I had that. On going inside to examine the moth, I found a large female

Eacles Imperialis, with not a scale of down misplaced. Even by gas light I could see that the yellow of the living moth was a warm canary colour, and the lavender of the mounted specimen closer heliotrope on the living, for there were pinkish tints that had faded from the pinned moth.

She was heavy with eggs, and made no attempt to fly, so I closed the box and left her until the lights were out, and then removed the lid. Every opening was tightly screened, and as she had mated, I did not think she would fly. I hoped in the freedom of the Cabin she would not break her wings, and ruin herself for a study.

There was much comfort in the thought that I could secure her likeness; her eggs would be fertile, and I could raise a brood the coming season, in which would be both male and female. When life was over I could add her to my specimen case, for these are of the moths that do not eat, and live only a few days after depositing their eggs. So I went out and explained to Mr. Pettis what efforts I had made to secure this yellow moth, comforted him for allowing the male to escape by telling him I could raise all I wanted from the eggs of the female, showed him my entire collection, and sent him from the Cabin such a friend to my work, that it was he who brought me an oil-coated lark a few days later.

On rising early the next morning, I found my moth had deposited some eggs on the dining-room floor, before the conservatory doors, more on the heavy tapestry that covered them, and she was clinging to a velvet curtain at a library window, liberally dotting it with eggs, almost as yellow as her body. I turned a tumbler over those on the floor, pinned folds in the curtains, and as soon as the light was good, set up a camera and focused on a suitable location.

She climbed on my finger when it was held before her, and was carried, with no effort to fly, to the place I had selected, though Molly-Cotton walked close with a spread net, ready for the slightest impulse toward movement. But female moths seldom fly until they have finished egg depositing, and this one was transferred with no trouble to the spot on which I had focused. On the back wall of the Cabin, among some wild roses, she was placed on a log, and immediately raised her wings, and started for the shade of the vines. The picture made of her as she walked is beautiful. After I had secured several studies she was returned to the library curtain, where she resumed egg placing. These were not counted, but there, were at least three hundred at a rough guess.

I had thought her lovely in gas light, but day brought forth marvels and wonders. When a child, I used to gather cowslips in a bed of lush swale, beside a little creek at the foot of a big hill on our farm. At the summit was an old orchard, and in a brush-heap a brown thrush nested. From a red winter pearmain the singer poured out his own heart in song, and then reproduced the love ecstasy of every other bird of the orchard. That moth's wings were so exactly the warm though delicate yellow of the flowers I loved, that as I looked at it I could feel my bare feet sinking in the damp ooze, smell the fragrance of the buttercups, and hear again the ripple of the water and the mating exultation of the brown thrush.

In the name—Eacles Imperialis—there is no meaning or appropriateness to "Eacles"; "Imperialis"—of course, translates imperial—which seems most fitting, for the moth is close the size of Cecropia, and of truly royal beauty. We called it the Yellow Emperor. Her Imperial Golden Majesty had a wing sweep of six and a quarter inches. From the shoulders spreading in an irregular patch over front and back wings, most on the front, were markings of heliotrope, quite dark in colour: Near the costa of the front wings were two almost circular dots of slightly paler heliotrope, the one nearest the edge about half the size of the other. On the back wings, halfway from each edge, and half an inch from the marking at the base, was one round spot of the same colour. Beginning at the apex of the front pair, and running to half an inch from the lower edge, was a band of escalloped heliotrope. On the back pair this band began half an inch from the edge and ran straight across, so that at the outer curve of the wing it was an inch higher. The front wing surface and the space above this marking on the back were liberally sprinkled with little oblong touches of heliotrope;

but from the curved line to the bases of the back pair, the colouring was pure canary yellow.

The top of the head was covered with long, silken hairs of heliotrope, then a band of yellow; the upper abdomen was strongly shaded with heliotrope almost to the extreme tip. The lower sides of the wings were yellow at the base, the spots showing through, but not the bands, and only the faintest touches of the mottling. The thorax and abdomen were yellow, and the legs heliotrope. The antennae were heliotrope, fine, threadlike, and closely pressed to the head. The eyes were smaller than those of Cecropia, and very close together.

Compared with Cecropia these moths were very easy to paint. Their markings were elaborate, but they could be followed accurately, and the ground work of colour was warm cowslip yellow. The only difficulty was to make the almost threadlike antennae show, and to blend the faint touches of heliotrope on the upper wings with the yellow.

The eggs on the floor and curtains were guarded with care. They were dotted around promiscuously, and at first were clear and of amber colour, but as the little caterpillars grew in them, they showed a red line three fourths of the way around the rim, and became slightly depressed in the middle. The young emerged in thirteen days. They were nearly half an inch long, and were yellow with black lines. They began the task of eating until they reached the pupa state, by turning on their shells and devouring all of them to the glue by which they were fastened.

They were given their choice of oak, alder, sumac, elm, cherry, and hickory. The majority of them seemed to prefer the hickory. They moulted on the fifth day for the first time, and changed to a brown colour. Every five or six days they repeated the process, growing larger and of stronger colour with each moult, and developing a covering of long white hairs. Part of these moulted four times, others five.

At past six weeks of age they were exactly as Mr. Eisen had described them to me. Those I kept in confinement pupated on a bed of baked gravel, in a tin bucket. It is imperative to bake any earth or sand used for them to kill pests invisible to the eye, that might bore into the pupa cases and destroy the moths.

I watched the transformation with intense interest. After the caterpillars had finished eating they travelled in search of a place to burrow for a day or two. Then they gave up, and lay quietly on the sand. The colour darkened hourly, the feet and claspers seemed to draw inside, and one morning on going to look there were some greenish brown pupae. They shone as if freshly varnished, as indeed they were, for the substance provided to facilitate the emergence of the pupae from the caterpillar skins dries in a coating, that helps to harden the cases and protect them. These pupae had burst the skins at the thorax, and escaped by working the abdomen until they lay an inch or so from the skins.

What a "cast off garment" those skins were! Only the frailest outside covering, complete in all parts, and rapidly turning to a dirty brown. The pupae were laid away in a large box having a glass lid. It was filled with baked sand, covered with sphagnum moss, slightly dampened occasionally, and placed where it was cool, but never at actual freezing point. The following spring after the delight of seeing them emerge, they were released, for I secured a male to complete my collection a few days later, and only grew the caterpillars to prove it possible.

There was a carnival in the village, and, for three nights the streets were illuminated brightly from end to end, to the height of Ferris wheels and diving towers. The lights must have shone against the sky for miles around, for they drew from the Limberlost, from the Canoper, from Rainbow Bottom, and the Valley of the Wood Robin, their winged creatures of night.

I know Emperors appear in these places in my locality, for the caterpillars feed on leaves found there, and enter the ground to pupate; so of course the moth of June begins its life in the same location. Mr. Pettis found the mated pair he brought to me, on a bush at the edge of a swamp. They also emerge in cities under any tree on which their caterpillars feed. Once late in May, in the corner of a lichen-covered, old snake fence beside the Wabash on the Shimp farm, I made a series of studies of the home life of a pair of ground sparrows. They had chosen for a location a slight depression covered with a rank growth of meadow grass. Overhead wild plum and thorn in full bloom lay white-sheeted against the blue sky; red bud spread its purple haze, and at a curve, the breast of the river gleamed white as ever woman's; while underfoot the grass was obscured with masses of wild flowers.

An unusually fine cluster of white violets attracted me as I worked around the birds, so on packing at the close of the day I lifted the plant to carry home for my wild flower bed. Below a few inches of rotting leaves and black mould I found a lively pupa of the Yellow Emperor.

So these moths emerge and deposit their eggs in the swamps, forests, beside the river and wherever the trees on which they feed grow. When the serious business of life is over, attracted by strong lights, they go with other pleasure seeking company, and grace society by their royal presence.

I could have had half a dozen fine Imperialis moths during the three nights of the carnival, and fluttering above buildings many more could be seen that did not descend to our reach. Raymond had such a busy time capturing moths he missed most of the joys of the carnival, but I truly think he liked the chase better. One he brought me, a female, was so especially large that I took her to the Cabin to be measured, and found her to be six and three quarter inches, and of the lightest yellow of any specimen I have seen. Her wings were quite ragged. I imagined she had finished laying her eggs, and was nearing the end of life, hence she was not so brilliant as a newly emerged specimen. The moth proved this theory correct by soon going out naturally.

Choice could be made in all that plethora, and a male and female of most perfect colouring and markings were selected, for my studies of a pair. One male was mounted and a very large female on account of her size. That completed my Imperialis records from eggs to caterpillars, pupae and moths.

The necessity for a book on this subject; made simple to the understanding, and attractive to the eye of the masses, never was so deeply impressed upon me as in an experience with Imperialis. Molly-Cotton was attending a house-party, and her host had chartered a pavilion at a city park for a summer night dance. At the close of one of the numbers; over the heads of the laughing crowd, there swept toward the light a large yellow moth.

With one dexterous sweep the host caught it, and while the dancers crowded around him with exclamations of wonder and delight, he presented it to Molly-Cotton and asked, "Do you know what it is?"

She laughingly answered, "Yes. But you don't!"

"Guilty!" he responded. "Name it."

For one fleeting instant Molly-Cotton measured the company. There was no one present who was not the graduate of a commissioned high school. There were girls who were students at The Castle, Smith, Vassar, and Bryn Mawr. The host was a Cornell junior, and there were men from Harvard and Yale.

"It is an Eacles Imperialis Io Polyphemus Cecropia Regalis," she said. Then in breathless suspense she waited.

"Shades of Homer!" cried the host. "Where did you learn it?"

"They are flying all through the Cabin at home," she replied. "There was a tumbler turned over their eggs on the dining-room floor, and you dared not sit on the right side of the library window seat because of them when I left."

"What do you want with their eggs?" asked a girl.

"Want to hatch their caterpillars, and raise them until they transform into these moths," answered poor Molly-Cotton, who had been taught to fear so few living things that at the age of four she had carried a garter snake into the house for a playmate.

"Caterpillars!" The chorus arose to a shriek. "Don't they sting you? Don't they bite you?"

"No, they don't!" replied Molly-Cotton. "They don't bite anything except leaves; they are fine big fellows; their colouring is exquisite; and they evolve these beautiful moths.

I invite all of you to visit us, and see for yourselves how intensely interesting they are."

There was a murmur of polite thanks from the girls, but one man measured Molly-Cotton from the top curl of her head to the tip of her slippers, and answered, "I accept the invitation. When may I come?" He came, and left as great a moth enthusiast as any of us. This incident will be recognized as furnishing the basis on which to build the ballroom scene in "A Girl of the Limberlost", in which Philip and Edith quarrel over the capture of a yellow Emperor. But what of these students from the great representative colleges of the United States, to whom a jumbled string made from the names, of half a dozen moths answered for one of the commonest of all?

CHAPTER V.
THE LADY BIRD:
DEILEPHILA LINEATA

In that same country garden where my first Cecropia was found, Deilephila Lineata was one of my earliest recollections. This moth flew among the flowers of especial sweetness all day long, just as did the hummingbirds; and I was taught that it was a bird also—the Lady Bird. The little tan and grey thing hovering in air before the flowers was almost as large as the humming-birds, sipping honey as they did, swift in flight as they; and both my parents thought it a bird.

They did not know the humming-birds were feasting on small insects attracted by the sweets, quite as often as on honey, for they never had examined closely. They had been taught, as I was, that this other constant visitor to the flowers was a bird. When a child, a humming-bird nested in a honeysuckle climbing over my mother's bedroom window. My father lifted me, with his handkerchief bound across my nose, on the supposition that the bird was so delicate it would desert its nest and eggs if they were breathed upon, to see the tiny cup of lichens, with a brown finish so fine it resembled the lining of a chestnut burr, and two tiny eggs. I well remember he told me that I now had seen the nest and eggs of the smallest feathered creature except the Lady Bird, and he never had found its cradle himself.

Every summer I discovered nests by the dozen, and for several years a systematic search was made for the home of a Lady Bird. One of the unfailing methods of finding locations was to climb a large Bartlett pear tree that stood beside the garden fence, and from an overhanging bough watch where birds flew with bugs and worms they collected. Lady Birds were spied upon, but when they left our garden they arose high in air, and went straight from sight toward every direction. So locating their nests as those of other birds were found, seemed impossible.

Then I tried going close the sweetest flowers, those oftenest visited, the petunias, yellow day lilies, and trumpet creepers, and sitting so immovably I was not noticeable while I made a study of the Lady Birds. My first discovery was that they had no tail. One poised near enough to make sure of that, and I hurried to my father with the startling news. He said it was nothing remarkable; birds frequently lost their tails. He explained how a bird in close quarters has power to relax its muscles, and let its tail go in order to save its body, when under the paw of a cat, or caught in a trap.

That was satisfactory, but I thought it must have been a spry cat to get even a paw on the Lady Bird, for frequently humming-birds could be seen perching, but never one of these. I watched the tail question sharply, and soon learned the cats had been after every Lady Bird that visited our garden, or any of our neighbours, for not one of them had a tail. When this information was carried my father, he became serious, but finally he said perhaps the tail was very short; those of humming-birds or wrens were, and apparently some water birds had no tail, or at least a very short one.

That seemed plausible, but still I watched this small and most interesting bird of all; this bird that no one ever had seen taking a bath, or perching, and whose nest never had been found by a person so familiar with all outdoors as my father. Then came a second discovery: it could curl its beak in a little coil when leaving a flower. A few days later I saw distinctly that it had four wings but I could discover no feet. I became a rank doubter, and when these convincing proofs were carried to my father, he also grew dubious.

"I always have thought and been taught that it was a bird," he said, "but you see so clearly and report so accurately, you almost convince me it is some large insect possibly of the moth family."

When I carried this opinion to my mother and told her, no doubt pompously, that 'very possibly' I had discovered that the Lady Bird was not a bird at all, she hailed it as high treason, and said, "Of course it is a bird!" That forced me to action. The desperate course of capturing one was resolved upon. If only I could, surely its feet, legs, and wings would tell if it were a bird. By the hour I slipped among those bloom-bordered walks between the beds of flaming sweet-williams, buttercups, phlox, tiger and day lilies, Job's tears, hollyhocks, petunias, poppies, mignonette, and every dear old-fashioned flower that grows, and followed around the flower-edged beds of lettuce, radishes, and small vegetables, relentlessly trailing Lady Birds.

Pass after pass I made at them, but they always dived and escaped me. At last, when I almost had given up the chase, one went nearly from sight in a trumpet creeper. With a sweep the flower was closed behind it, and I ran into the house crying that at last I had caught a Lady Bird. Holding carefully, the trumpet was cut open with a pin, and although the moth must have been slightly pinched, and lacking in down when released, I clung to it until my mother and every doubting member of my family was convinced that this was no bird at all, for it lacked beak, tail, and feathers, while it had six legs and four wings. Father was delighted that I had learned something new, all by myself; but I really think it slightly provoked my mother when thereafter I always refused to call it a bird. This certainly was reprehensible. She should have known all the time that it was a moth.

The other day a club woman of Chicago who never in her life has considered money, who always has had unlimited opportunities for culture both in America and Europe, who speaks half a dozen languages, and has the care of but one child, came in her auto mobile to investigate the Limberlost.

Almost her first demand was to see pictures. One bird study I handed her was of a brooding king rail, over a foot tall, with a three-foot wing sweep, and a long curved bill. She cried, "Oh! see the dear little hummingbird!"

If a woman of unlimited opportunity, in this day of the world, does not know a rail from a humming-bird, what could you expect of my little mother, who spoke only two languages, reared twelve lusty children, and never saw an ocean.

So by degrees the Lady Bird of the garden resolved itself into Deilephila Lineata. Deile—evening; phila—lover; lineata—lined; the Lined Evening Lover. Why 'evening' is difficult to understand, for all my life this moth occurs more frequently with me in the fore and early afternoon than in the evening. So I agree with those entomologists who call it the 'white-lined morning-sphinx.' It is lovely in modest garb, delicately lined, but exceedingly rich in colour. It has the long slender wings of the Sphingid moths, and in grace and tirelessness of flight resembles Celeus, the swallow of the moth family.

Its head is very small, and its thorax large. The eyes are big, and appear bigger because set in so tiny a head. Under its tongue, which is a full inch long, is a small white spot that divides, spreads across each eye, and runs over the back until even with the bases of the front wings. The top of the head and shoulders are olive brown, decorated with one long white line dividing it in the middle, and a shorter on each side. The abdomen is a pale brown, has a straight line running down the middle of the back, made up of small broken squares of very dark brown, touched with a tiny mark of white. Down each side of this small line extends a larger one, wider at the top and tapering, and this is composed of squares of blackish brown alternating with white, the brown being twice the size of the white. The sides of the abdomen are flushed with beautiful rosy pink, and beneath it is tan colour.

The wings are works of art. The front are a rich olive brown, marked the long way in the middle by a wide band of buff, shading to lighter buff at the base. They are edged from the costa to where they meet the back wings, with a line of almost equal width of darker buff, the lower edge touched with white. Beginning at the base, and running an equal distance apart from the costa to this line, are fine markings of white, even and clear as if laid on with a ruler.

The surprise comes in the back wings, that show almost entirely when the moth is poised before a flower. These have a small triangle of the rich dark brown, and a band of the same at the lower edge, with a finish of olive, and a fine line of white as a marginal decoration. Crossing each back wing is a broad band of lovely pink of deeper shade than the colour on the sides. This pink, combined with the olive, dark browns, and white lining, makes the colour scheme of peculiar richness.

Its antennae are long, clubbed, and touched with white at the tips. The legs and body are tan colour. The undersides of the wings are the same as the upper, but the markings of brown and buffish pink show through in lighter colour, while the white lining resembles rows of tan ridges beneath. Its body is covered with silky hairs, longest on the shoulders, and at the base of the wings.

The eggs of the moth are laid on apple, plum, or woodbine leaves, or on grape, currant, gooseberry, chickweed or dock. During May and June around old log cabins in the country, with gardens that contain many of these vines and bushes, and orchards of bloom where the others can be found the Lined Evening Lover deposits her eggs.

The caterpillars emerge in about six days. The tiny ovoid eggs are a greenish yellow. The youngsters are pale green, and have small horns. After a month spent in eating, and skin casting, the full-grown caterpillar is over two inches long, and as a rule a light green. There are on each segment black patches, that have a touch of orange, and on that a hint of yellow. The horn increases with the growth of the caterpillar, can be moved at will, and seems as if it were a vicious 'stinger.' But there is no sting, or any other method of self-defence, unless the habit of raising the head and throwing it from side to side could be so considered. With many people, this movement, combined with the sharp horn, is enough, but as is true of most caterpillars, they are perfectly harmless.

Some moth historians record a mustard yellow caterpillar of this family, and I remember having seen some that answer the description; but all I ever have known to be Lineata were green.

The pupae are nearly two inches long and are tan coloured. They usually are found in the ground in freedom, or deep under old logs among a mass of leaves spun together. In captivity the caterpillars seem to thrive best on a diet of purslane, and they pupate perfectly on dry sand in boxes.

These moths have more complete internal development than those of night, for they feed and live throughout the summer. I photographed a free one feasting on the sweets of petunias in a flower bed at the Cabin, on the seventh of October.

CHAPTER VI.
MOTHS OF THE MOON:
ACTIAS LUNA

One morning there was a tap at my door, and when I opened it I found a tall, slender woman having big, soft brown eyes, and a winning smile. In one hand she held a shoe-box, having many rough perforations. I always have been glad that my eyes softened at the touch of pleading on her face, and a smile sprang in answer to hers before I saw what she carried. For confession must be made that a perforated box is a passport to my good graces any day.

The most wonderful things come from those that are brought to my front door. Sometimes they contain a belated hummingbird, chilled with the first heavy frost of autumn, or a wounded weasel caught in a trap set for it near a chicken coop, or a family of baby birds whose parents some vandal has killed. Again they carry a sick or wounded bird that I am expected to doctor; and butterflies, moths, insects, and caterpillars of every description.

"I guess I won't stop," said the woman in answer to my invitation to enter the Cabin. "I found this creature on my front porch early this morning, and I sort of wanted to know what it was, for one thing, and I thought you might like to have it, for another."

"Then of course you will come in, and we will see what it is," I answered, leading the way into the library.

There I lifted the lid slightly to take a peep, and then with a cry of joy, opened it wide. That particular shoe-box had brought me an Actias Luna, newly emerged, and as yet unable to fly. I held down my finger, it climbed on, and was lifted to the light.

"Ain't it the prettiest thing?" asked the woman, with stars sparkling in her dark eyes. "Did you ever see whiter white?"

Together we studied that moth. Clinging to my finger, the living creature was of such delicate beauty as to impoverish my stock of adjectives at the beginning. Its big, pursy body was covered with long, furry scales of the purest white imaginable. The wings were of an exquisite light green colour; the front pair having a heavy costa of light purple that reached across the back of the head: the back pair ended in long artistic 'trailers,' faintly edged with light yellow. The front wing had an oval transparent mark close the costa, attached to it with a purple line, and the back had circles of the same. These decorations were bordered with lines of white, black, and red. At the bases of the wings were long, snowy silken hairs; the legs were purple, and the antennae resembled small, tan-coloured ferns. That is the best I can do at description. A living moth must be seen to form a realizing sense of its shape and delicacy of colour. Luna is our only large moth having trailers, and these are much longer in proportion to size and of more graceful curves than our trailed butterflies.

The moth's wings were fully expanded, and it was beginning to exercise, so a camera was set up hastily, and several pictures of it secured. The woman helped me through the entire process, and in talking with her, I learned that she was Mrs. McCollum, from a village a mile and a half north of ours; that when she reached home she would have walked three miles to make the trip; and all her neighbours had advised her not to come, but she "had a feeling that she would like to."

"Are you sorry?" I asked.

"Am I sorry!" she cried. "Why I never had a better time in my life, and I can teach the children what you have told me. I'll bring you everything I can get my fingers on that you can use, and send for you when I find bird nests."

Mrs. McCollum has kept that promise faithfully. Again and again she trudged those three miles, bringing me small specimens of many species or to let me know that she had found a nest.

A big oak tree in Mrs. McCollum's yard explained the presence of a Luna there, as the caterpillars of this specie greatly prefer these leaves. Because the oak is of such slow growth it is seldom planted around residences for ornamental purposes; but is to be found most frequently in the forest. For this reason Luna as a rule is a moth of the deep wood, and so is seldom seen close a residence, making people believe it quite rare. As a matter of fact, it is as numerous where the trees its caterpillars frequent are to be found, as any other moth in its natural location. Because it is of the forest, the brightest light there is to attract it is the glare of the moon as it is reflected on the face of a murky pool, or on the breast of the stream rippling its way through impassable thickets. There must be a self-satisfied smile on the face of the man in the moon, in whose honour these delicate creatures are named, when on fragile wing they hover above his mirrored reflection; for of all the beauties of a June night in the forest, these moths are most truly his.

In August of the same year, while driving on a corduroy road in Michigan, I espied a Luna moth on the trunk of a walnut tree close the road. The cold damp location must account for this late emergence; for subsequent events proved that others of the family were as slow in appearing. A storm of protest arose, when I stopped the carriage and started to enter the swamp. The remaining occupants put in their time telling blood-curdling experiences with 'massaugers,' that infested those marshes; and while I bent grasses and cattails to make the best footing as I worked my way toward the moth, I could hear a mixed chorus "brought up thirteen in the dredge at the cement factory the other day," "killed nine in a hayfield below the cemetery," "saw a buster crossing the road before me, and my horse almost plunged into the swamp," "died of a bite from one that struck him while fixing a loose board in his front walk."

I am dreadfully afraid of snakes, and when it seemed I could not force myself to take another step, and I was clinging to a button bush while the water arose above my low shoes, the moth lowered its wings flat against the bark. From the size of the abdomen I could see that it was a female heavily weighted with eggs. Possibly she had mated the previous night, and if I could secure her, Luna life history would be mine.

So I set my teeth and advanced. My shoes were spoiled, and my skirts bedraggled, but I captured the moth and saw no indication of snakes. Soon after she was placed in a big pasteboard box and began dotting eggs in straight lines over the interior. They were white but changed colour as the caterpillars approached time to hatch. The little yellow-green creatures, nearly a quarter of an inch long, with a black line across the head, emerged in about sixteen days, and fed with most satisfaction on oak, but they would take hickory, walnut or willow leaves also. When the weather is cold the young develop slower, and I have had the egg period stretched to three weeks at times. Every few days the young caterpillars cast their skins and emerged in brighter colour and larger in size. It is usually supposed they mature in four moults, and many of them do, but some cast a fifth skin before transforming. When between seven and eight weeks of age, they were three inches long, and of strong blue-green colour. Most of them had tubercles of yellow, tipped with blue, and some had red.

They spun a leaf-cover cocoon, much the size and shape of that of Polyphemus, but whiter, very thin, with no inner case, and against some solid surface whenever possible. Fearing I might not handle them rightly, and lose some when ready to spin, I put half on our walnut tree so they could weave their cocoons according to characteristics.

They are fine, large, gaudy caterpillars. The handsomest one I ever saw I found among some gifts offered by Molly-Cotton for the celebration of my birthday. It had finished feeding, soon pupated in a sand pail and the following spring a big female emerged that attracted several males and they posed on a walnut trunk for beautiful studies.

Once under the oak trees of a summer resort, Miss Katherine Howell, of Philadelphia, intercepted a Luna caterpillar in the preliminary race before pupation and brought it to me. We offered young oak leaves, but they were refused, so it went before the camera. Behind the hotel I found an empty hominy can in which it soon began spinning, but it seemed to be difficult to fasten the threads to the tin, so a piece of board was cut and firmly wedged inside. The caterpillar clung to this and in the darkness of the can spun the largest and handsomest Luna winter quarters of all my experience.

Luna hunters can secure material from which to learn this exquisite creature of night, by searching for the moths on the trunks of oak, walnut, hickory, birch or willow, during the month of June. The moths emerge on the ground, and climb these trees to unfold and harden their wings. The females usually remain where they are, and the males are attracted to them. If undisturbed they do not fly until after mating and egg depositing are accomplished. The males take wing as soon as dusk of the first night arrives, after their wings are matured. They usually find the females by ten o'clock or midnight, and remain with them until morning. I have found mated pairs as late as ten o'clock in the forenoon.

The moths do not eat, and after the affairs of life are accomplished, they remain in the densest shade they can find for a few days, and fly at night, ending their life period in from three days to a week. Few of these gaudily painted ones have the chance to die naturally, for both birds and squirrels prey upon them, tearing away the delicate wings, and feasting on the big pulpy bodies.

White eggs on the upper side of leaves of the trees mentioned are a sign of Luna caterpillars in deep woods, and full-grown larvae can be found on these trees in August. By breaking off a twig on which they are feeding, carrying them carefully, placing them in a box where they cannot be preyed upon by flies and parasites, and keeping a liberal supply of fresh damp leaves, they will finish the feeding days, and weave their cocoons.

Or the cocoons frequently can be found already spun among the leaves, by nutting parties later in the fall. There is small question if Luna pupae be alive, for on touching the cocoons they squirm and twist so vigorously that they can be heard plainly. There is so little difference in the size of male and female Lunas, that I am not sure of telling them apart in the cocoon, as I am certain I can Cecropia.

Cocoon gathering in the fall is one of the most delightful occupations imaginable. When flowers are gone; when birds have migrated; when brilliant foliage piles knee deep underfoot; during those last few days of summer, zest can be added to a ramble by a search for cocoons. Carrying them home with extreme care not to jar or dent them, they are placed in the conservatory among the flowers. They hang from cacti spines and over thorns on the big century plant and lemon tree. When sprinkling, the hose is turned on them, as they would take the rain outside. Usually they are placed in the coolest spots, where ventilation is good.

There is no harm whatever in taking them *if the work is carefully and judiciously done.* With you they are safe. Outside they have precarious chance for existence, for they are constantly sought by hungry squirrels and field mice, while the sharp eyes and sharper beaks of jays, and crows, are for ever searching for them. The only danger is in keeping them too warm, and so causing their emergence before they can be placed out safely at night, after you have made yourself acquainted with Luna history.

If they are kept cool enough that they do not emerge until May or June, then you have one of the most exquisite treats nature has in store for you, in watching the damp spot spread on the top of the cocoon where an acid is ejected that cuts and softens the tough fibre, and allows the moth to come pushing through in the full glory of its gorgeous birth. Nowhere in nature can you find such delicate and daintily shaded markings or colours so brilliant and fresh as on the wings of these creatures of night.

After you have learned the markings and colours, and secured pictures if you desire, and they begin to exhibit a restlessness, as soon as it is dusk, release them. They are as well prepared for all life has for them as if they had emerged in the woods. The chances are that they are surer of life at your hands than they would have been if left afield, provided you keep them cool enough that they do not emerge too soon. If you want to photograph them, do it when the wings are fully developed, but before they have flown. They need not be handled; their wings are unbroken; their down covering in place to the last scale; their colours never so brilliant; their markings the plainest they ever will be; their big pursy bodies full of life; and they will climb with perfect confidence on any stick, twig, or limb held before them. Reproductions of them are even more beautiful than those of birds. By all means photograph them out of doors on a twig or leaf that their caterpillars will eat. Moths strengthen and dry very quickly outside in the warm crisp air of May or June, so it is necessary to have some one beside you with a spread net covering them, in case they want to fly before you are ready to make an exposure. In painting this moth the colours always should be copied from a living specimen as soon as it is dry. No other moth of my acquaintance fades so rapidly.

Repeatedly I am asked which I think the most beautiful of these big night moths. I do not know. All of them are indescribably attractive. Whether a pale green moth with purple markings is lovelier than a light yellow moth with heliotrope decorations; or a tan and brown one with pink lines, is a difficult thing to determine. When their descriptions are mastered, and the colour combinations understood, I fancy each person will find the one bearing most of his favourite colour the loveliest. It may be that on account of its artistically cut and coloured trailers, Luna has a touch of grace above any.

CHAPTER VII.
KING OF THE HOLLYHOCKS:
PROTOPARCE CELEUS

Protoparce Celeus was the companion of Deilephila Lineata in the country garden where I first studied Nature. Why I was taught that Lineata was a bird, and Celeus a moth, it is difficult to understand, for they appear very similar when poising before flowers. They visit the same blooms, and vary but little in size. The distinction that must have made the difference was that while Lineata kept company with the hummingbirds and fed all day, Celeus came forth at dusk, and flew in the evening and at night. But that did not conclusively prove it a moth, for nighthawks and whip-poor-wills did the same; yet unquestionably they were birds.

Anyway, I always knew Celeus was a moth, and that every big, green caterpillar killed on the tomato vines meant one less of its kind among the flowers. I never saw one of these moths close a tomato or potato vine, a jimson weed or ground cherry, but all my life I have seen their eggs on these plants, first of a pale green closely resembling the under side of the leaves, and if they had been laid some time, a yellow colour. The eggs are not dotted along in lines, or closely placed, but are deposited singly, or by twos, at least very sparsely.

The little caterpillars emerge in about a week, and then comes the process of eating until they grow into the large, green tomato or tobacco worms that all of us have seen. When hatched the caterpillars are green, and have grey caudal horns similar to Lineata. After eating for four or five days, they cast their skins. This process is repeated three or four times, when the full-grown caterpillars are over four inches long, exactly the colour of a green tomato, with pale blue and yellow markings of beautiful shades, the horns blue-black; and appearing sharp enough to inflict a severe wound.

Like all sphinx caterpillars Celeus is perfectly harmless; but this horn, in connexion with the habit the creatures have of clinging to the vines with the back feet, raising the head and striking from side to side, makes people very sure they can bite or sting, or inflict some serious hurt. So very vigorous are they in self-defence when disturbed, that robins and cuckoos are the only birds I ever have seen brave enough to pick them until the caterpillars loosen their hold and drop to the ground, where they are eaten with evident relish.

One cuckoo of my experience that nested in an old orchard, adjoining a potato patch, frequently went there caterpillar-hunting, and played havoc with one wherever found. The shy, deep wood habits of the cuckoo prevent it from coming close houses and into gardens, but robins will take these big caterpillars from tomato vines. However, they go about it rather gingerly, and the work of reducing one to non-resistance does not seem to be at all coveted. Most people exhibit symptoms of convulsions at sight of one. Yet it is a matter of education. I have seen women kiss and fondle cats and dogs, one snap from which would result in disfiguration or horrible death, and seem not to be able to get enough of them. But they were quite equal to a genuine faint if contact were suggested with a perfectly harmless caterpillar, a creature lacking all means of defence, save this demonstration of throwing the head.

When full-fed the caterpillars enter the earth to pupate, and on the fifteenth of October, 1906, only the day before I began this chapter, the Deacon, in digging worms for a fishing trip to the river, found a pupa case a yard from the tomato vines, and six inches below the surface. He came to my desk, carrying on a spade a ball of damp earth larger than a quart bowl. With all care we broke this as nearly in halves as possible and found in the centre a firm, oval hole, the size and shape of a hen's egg, and in the opening a fine fresh pupa case.

It was a beautiful red-brown in colour, long and slenderer than a number of others in my box of sand, and had a long tongue case turned under and fastened to the pupa between the wing shields. The sides of the abdomen were pitted; the shape of the head, and the eyes showed through the case, the wing shields were plainly indicated, and the abdominal shield was in round sections so that the pupa could twist from side to sid when touched, proving that the developing moth inside was very much alive and in fine condition.

There were no traces of the cast skin. The caterpillar had been so strong and had pushed so hard against the surrounding earth that the direction from which it had entered was lost. The soil was packed and crowded firmly for such a distance that this large ball was forced together. Trembling with eagerness I hurriedly set up a camera. This phase of moth life often has been described, but I never before heard of any one having been able to reproduce it, so my luck was glorious. A careful study of this ball of earth, the opening in which the case lies, and the pupa, with its blunt head and elaborate tongue shield, will convince any one that when ready to emerge these moths must bore the six inches to the surface with the point of the abdomen, and there burst the case, cling to the first twig and develop and harden the wings. The abdominal point is sharp, surprisingly strong, and the rings of the segments enable it to turn in all directions, while the earth is mellow and moist with spring rains. To force a way head first would be impossible on account of the delicate tongue shield, and for the moth to emerge underground and dig to the surface without displacing a feather of down, either before or after wing expansion, is unthinkable. Yet I always had been in doubt as to precisely how the exit of a pupa case moth took place, until I actually saw the earth move and the sharp abdominal point appear while working in my garden.

Living pupae can be had in the fall, by turning a few shovels of soil close vegetables in any country garden. In the mellow mould, among cabbages and tomato vines, around old log cabins close the Limberlost swamp, they are numerous, and the emerging moths haunt the sweet old-fashioned flowers.

The moth named Celeus, after a king of Eleusis, certainly has kingly qualities to justify the appellation. The colouring is all grey, black, brown, white and yellow, and the combinations are most artistic. It is a relative of Lineata. It flies and feeds by day, has nearly the same length of life, and is much the same in shape.

The head is small and sharp, eyes very much larger than Lineata, and tongue nearly four inches in length. The antennae are not clubbed, but long and hairlike. It has the broad shoulders, the long wings, and the same shape of abdomen. The wings, front and back, are so mottled, lined, and touched with grey, black, brown and white, as to be almost past definite description. The back wings have the black and white markings more clearly defined. The head meets the thorax with a black band. The back is covered with long, grey down, and joins the abdomen, with a band of black about a quarter of an inch wide, and then a white one of equal width. The abdomen is the gaudiest part of the moth. In general it is a soft grey. It is crossed by five narrow white lines the length of the abdomen, and a narrow black one down the middle. Along each side runs a band of white. On this are placed four large yellow spots each circled by a band of black that joins the black band of the spot next to it. The legs and under side of the abdomen and wings are a light grey-tan, with the wing markings showing faintly, and the abdomen below is decorated with two small black dots.

My first Celeus, a very large and beautiful one, was brought to me by Mr. Wallace Hardison, who has been an interested helper with this book. The moth had a wing sweep of fully five and a half inches, and its markings were unusually bright and strong. No other Celeus quite so big and beautiful ever has come to my notice. From four and a half to five inches is the average size.

There was something the matter with this moth. Not a scale of down seemed to be missing, but it was torpid and would not fly. Possibly it had been stung by some parasite before taking flight at all, for it was very fresh. I just had returned from a trip north, and there were some large pieces of birch bark lying on the table on which the moth had been placed. It climbed on one of these, and clung there, so I set up the bark, and made a time exposure. It felt so badly it did not even close them when I took a brush and spread its wings full width. Soon after it became motionless. I had begun photographing moths recently; it was one of my very first, and no thought of using it for natural history purposes occurred at the time. I merely made what I considered a beautiful likeness, and this was so appreciated whenever shown, that I went further and painted it in water colours.

Since moth pictures have accumulated, and moth history has engrossed me with its intense interest, I have been very careful in making studies to give each one its proper environment when placing it before my camera. Of all the flowers in our garden, Celeus prefers the hollyhocks. At least it comes to them oftenest and remains at them longest. But it moves continually and flies so late that a picture of it has been a task. After years of fruitless effort, I made one passable snapshot early in July, while the light was sufficiently strong that a printable picture could be had by intensifying the plate, and one good time exposure as a Celeus, with half-folded wings, clambered over a hollyhock, possibly hunting a spot on which to deposit an egg or two. The hollyhock painting of this chapter is from this study. The flowers were easy but it required a second trial to do justice to the complicated markings of the moth.

This evening lover and strong flyer, with its swallow-like sweep of wing, comes into the colour schemes of nature with the otter, that at rare times thrusts a sleek grey head from the river, with the grey-brown cotton-tails that bound across the stubble, and the coots that herald dawn in the marshes. Exactly the shades, and almost the markings of its wings can be found on very old rail fences. This lint shows lighter colour, and even grey when used in the house building of wasps and orioles, but I know places in the country where I could carve an almost perfectly shaded Celeus wing from a weather-beaten old snake fence rail.

Celeus visits many flowers, almost all of the trumpet-shaped ones, in fact, but if I were an artist I scarcely would think it right to paint a hollyhock without putting King Celeus somewhere in the picture, poised on his throne of air before a perfect bloom as he feasts on pollen and honey. The holly-hock is a kingly flower, with its regally lifted heads of bright bloom, and that the king of moths should show his preference for it seems eminently fitting, so we of the Cabin named him King of the Hollyhocks.

CHAPTER VIII.
HERA OF THE CORN:
HYPERCHIRA IO

At the same time he gave me the Eacles Imperialis moths, Mr. Eisen presented me with a pair of Hyperchiria Io. They were nicely mounted on the black velvet lining of a large case in my room, but I did not care for them in the least. A picture I would use could not be made from dead, dried specimens, and history learned from books is not worth knowing, in comparison with going afield and threshing it out for yourself in your own way. Because the Io was yellow, I wanted it—more than several specimens I had not found as yet, for yellow, be it on the face of a flower, on the breast of a bird, or in the gold of sunshine, always warms the depths of my heart.

One night in June, sitting with a party of friends in the library, a shadow seemed to sweep across a large window in front. I glanced up, and arose with a cry that must have made those present doubt my sanity. A perfect and beautiful Io was walking leisurely across the glass.

"A moth!" I cried. "I have none like it! Deacon, get the net!"

I caught a hat from the couch, and ran to the veranda. The Deacon followed with the net.

"I was afraid to wait," I explained. "Please bring a piece of pasteboard, the size of this brim."

I held the hat while the Deacon brought the board. Then with trembling care we slipped it under, and carefully carried the moth into the conservatory. First we turned on the light, and made sure that every ventilator was closed; then we released the Io for the night. In the morning we found a female clinging to a shelf, dotting it with little top-shaped eggs. I was delighted, for I thought this meant the complete history of a beautiful moth. So exquisite was the living, breathing creature, she put to shame the form and colouring of the mounted specimens. No wonder I had not cared for them!

Her fore-wings were a strong purplish brown in general effect, but on close examination one found the purplish tinge a commingling of every delicate tint of lavender and heliotrope imaginable. They were crossed by escalloped bands of greyish white, and flecked with touches of the same, seeming as if they had been placed with a brush. The back wings were a strong yellow. Each had, for its size, an immense black eye-spot, with a blue pupil covering three-fourths of it, crossed by a perfect comma of white, the heads toward the front wings and the curves bending outward. Each eye-spot was in a yellow field, strongly circled with a sharp black line; then a quarter of an inch band of yellow; next a heliotrope circle of equal width; yellow again twice as wide;

then a faint heliotrope line; and last a very narrow edging of white. Both wings joined the body under a covering of long, silky, purple-brown hairs.

She was very busy with egg depositing, and climbed to the twig held before her without offering to fly. The camera was carried to the open, set up and focused on a favourable spot, while Molly-Cotton walked beside me holding a net over the moth in case she took flight in outer air. The twig was placed where she would be in the deepest shade possible while I worked rapidly with the camera.

By this time experience had taught me that these creatures of moonlight and darkness dislike the open glare of day, and if placed in sunlight will take flight in search of shade more quickly than they will move if touched. So until my Io settled where I wanted her with the wings open, she was kept in the shadow. Only when I grasped the bulb and stood ready to snap, was the covering lifted, and for the smallest fraction of a second the full light fell on her; then darkness again.

In three days it began to be apparent there was something wrong with the eggs. In four it was evident, and by five I was not expecting the little caterpillars to emerge, and they did not. The moth had not mated and the eggs were not fertile. Then I saw my mistake. Instead of shutting the female in the conservatory at night, I should have tied a soft cotton string firmly around her body, and fastened it to some of the vines on the veranda. Beyond all doubt, before morning, a male of her kind would have been attracted to her.

One learns almost as much by his mistakes as he profits by his successes in this world. Writing of this piece of stupidity, at a time in my work with moths when a little thought would have taught me better, reminds me of an experience I had with a caterpillar, the first one I ever carried home and tried to feed. I had an order to fill for some swamp pictures, and was working almost waist deep in a pool in the Limberlost, when on a wild grape-vine swinging close to my face, I noticed a big caterpillar placidly eating his way around a grape leaf. The caterpillar was over four inches long, had no horn, and was of a clear red wine colour, that was beautiful in the sunlight. I never before had seen a moth caterpillar that was red and I decided it must be rare. As there was a wild grapevine growing over the east side of the Cabin, and another on the windmill, food of the right kind would be plentiful, so I instantly decided to take the caterpillar home. It was of the specimens that I consider have almost 'thrust themselves upon me.'

When the pictures were finished and my camera carried from the swamp, I returned with the clippers and cut off vine and caterpillar, to carry with me. On arrival I placed it in a large box with sand on the bottom, and every few hours took out the wilted leaves, put in fresh ones, and sprinkled them to insure crispness, and to give a touch of moisture to the atmosphere in the box, that would make it seem more like the swamp.

My specimen was readily identified as Philampelus Pandorus, of which I had no moth, so I took extra care of it in the hope of a new picture in the spring. It had a little flat head that could be drawn inside the body like a turtle, and on the sides were oblique touches of salmon. Something that appeared to be a place for a horn could be seen, and a yellow tubercle was surrounded by a black line. It ate for three days, and then began racing so frantically around the box, I thought confinement must be harmful, so I gave it the freedom of the Cabin, warning all my family to 'look well to their footsteps.' It stopped travelling after a day or two at a screen covering the music-room window, and there I found it one morning lying still, a shrivelled, shrunken thing; only half the former length, so it was carefully picked up, and thrown away!

Of course the caterpillar was in the process of changing into the pupa, and if I had known enough to lay it on the sand in my box, and wait a few days, without doubt a fine pupa would have emerged from that shrunken skin, from which, in the spring, I could have secured an exquisite moth, with shades of olive green, flushed with pink. The thought of it makes me want to hide my head. It was six years before I found a living moth, or saw another caterpillar of that species.

A few days later, while watching with a camera focused on the nest of a blackbird in Mrs. Corson's woods east of town, Raymond, who was assisting me, crept to my side and asked if it would do any harm for him to go specimen hunting. The long waits with set cameras were extremely tedious to the restless spirits of the boy, and the birds were quite tame, the light was under a cloud, and the woods were so deep that after he had gone a few rods he was from sight, and under cover; besides it was great hunting ground, so I gladly told him to go.

The place was almost 'virgin,' much of it impassable and fully half of it was under water that lay in deep, murky pools throughout summer. In the heat of late June everything was steaming; insect life of all kinds was swarming; not far away I could hear sounds of trouble between the crow and hawk tribes; and overhead a pair of black vultures, whose young lay in a big stump in the interior, were searching for signs of food. If ever there was a likely place for specimens it was here; Raymond was an expert at locating them, and fearless to foolhardiness. He had been gone only a short time when I heard a cry, and I knew it must mean something, in his opinion, of more importance than blackbirds.

I answered "Coming," and hastily winding the long hose, I started in the direction Raymond had taken, calling occasionally to make sure I was going the right way. When I found him, the boy was standing beside a stout weed, hat in hand, intently watching something. As I leaned forward I saw that it was a Hyperchiria Io that just had emerged from the cocoon, and as yet was resting with wings untried. It differed so widely from my moth of a few days before, I knew it must be a male.

This was only three-fourths as large as mine, but infinitely surpassed it in beauty. Its front wings were orange-yellow, flushed with red-purple at the base, and had a small irregular brown spot near the costa. Contrary to all precedent, the under side of these wings were the most beautiful, and bore the decorations that, in all previous experience with moths, had been on the upper surface, faintly showing on the under. For instance, this irregular brown marking on the upper side proved to be a good-sized black spot with with white dot in the middle on the under; and there was a curved line of red-purple from the apex of the wing sloping to the lower edge, nearly half an inch from the margin. The space from this line to the base of the wing was covered with red-purple down. The back wings were similar to the female's, only of stronger colour, and more distinct markings; the eye-spot and lining appeared as if they had been tinted with strong fresh paint, while the edges of the wings lying beside the abdomen had the long, silken hairs of a pure, beautiful red their entire length:

A few rods away men were ploughing in the adjoining corn field, and I remembered that the caterpillar of this moth liked to feed on corn blades, and last summer undoubtedly lived in that very field. When I studied Io history in my moth books, I learned these caterpillars ate willow, wild cherry, hickory, plum, oak, sassafras, ash, and poplar. The caterpillar was green, more like the spiny butterfly caterpillars than any moth one I know. It had brown and white bands, brown patches, and was covered with tufts of stiff upstanding spines that pierced like sharp needles. This was not because the caterpillar tried to hurt you, but because the spines were on it, and so arranged that if pressed against, an acid secretion sprang from their base. This spread over the flesh the spines touched, stinging for an hour like smartweed, or nettles.

When I identified this caterpillar in my books, it came to me that I had known and experienced its touch. But it did not forcibly impress me until that instant that I knew it best of all, and that it was my childhood enemy of the corn. Its habit was to feed on the young blades, and cling to them with all its might. If I was playing Indian among the rows, or hunting an ear with especially long, fine 'silk' for a make-believe doll, or helping the cook select ears of Jersey Sweet to boil for dinner, and accidentally brushed one of these caterpillars with cheek or hand, I felt its burning sting long afterward. So I disliked those caterpillars.

For I always had played among the corn. Untold miles I have ridden the plough horses across the spring fields, where mellow mould rolled black from the shining shares, and the perfumed air made me feel so near flying that all I seemed to need was a high start to be able to sail with the sentinel blackbird, that perched on the big oak, and with one sharp 'T'check!' warned his feeding flock, surely and truly, whether a passing man carried a gun or a hoe. Then came the planting, when bare feet loved the cool earth, and trotted over other untold miles, while little fingers carefully counted out seven grains from the store carried in my apron skirt, as I chanted:

"One for the blackbird, one for the crow; One for the cutworm and four to grow."

Then father covered them to the right depth, and stamped each hill with the flat of the hoe, while we talked of golden corn bread, and slices of mush, fried to a crisp brown that cook would make in the fall. We had to plant enough more to feed all the horses, cattle, pigs, turkeys, geese, and chickens, during the long winter, even if the sun grew uncomfortably warm, and the dinner bell was slow about ringing.

Then there were the Indian days in the field, when a fallen eagle feather stuck in a braid, and some pokeberry juice on the face, transformed me into the Indian Big Foot, and I fled down green aisles of the corn before the wrath of the mighty Adam Poe. At times Big Foot grew tired fleeing, and said so in remarkably distinct English, and then to keep the game going, my sister Ada, who played Adam Poe, had to turn and do the fleeing or be tomahawked with a stick.

When the milk was in the ears, they were delicious steamed over salted water, or better yet roasted before coals at the front of the cooking stove, and eaten with butter and salt, if you have missed the flavour of it in that form, really you never have known corn!

Next came the cutting days. These were after all the caterpillars had climbed down, and travelled across the fence to spin their cocoons among the leaves of the woods; as if some instinct warned them that they would be ploughed up too early to emerge, if they remained in the field. The boys bent four hills, lashed the tassels together for a foundation, and then with one sweep of their knives, they cut a hill at a time, and stacked it in large shocks, that lined the field like rows of sentinels, guarding the gold of pumpkin and squash lying all around. While the shocks were drying, the squirrels, crows, and quail took possession, and fattened their sides against snow time.

Then the gathering days of October—they were the best days of all! Like a bloom-outlined vegetable bed, the goldenrod and ironwort, in gaudy border, filled the fence corners of the big fields. A misty haze hung in the air, because the Indians were burning the prairies to round up game for winter. The cawing of the crows, the chatter of blackbirds, and the piping bob-whites, sounded so close and so natural out there, while the crowing cocks of the barnyard seemed miles away and slightly unreal. Grown up and important, I sat on a board laid across the wagon bed, and guided the team of matched greys between the rows of shocks, and around the 'pie-timber' as my brother Leander called the pumpkins while father and the boys opened the shocks and husked the ears. How the squirrels scampered to the woods and to the business of storing away the hickory nuts that we could hear rattling down every frosty morning! We hurried with the corn; because as soon as the last shock was in, we might take the horses, wagon, and our dinner, and go all day to the woods, where we gathered our winter store of nuts. Leander would take a gun along, and shoot one of those saucy squirrels for the little sick mother.

Last came the November night, when the cold had shut us in. Then selected ears that had been dried in the garret were brought down, white for 'rivel' and to roll things in to fry, and yellow for corn bread and mush. A tub full of each was shelled, and sacked to carry to the mill the following day. I sat on the floor while father and the boys worked, listening to their talk, as I built corncob castles so high they toppled from their many stories. Sometimes father made cornstock fiddles that would play a real tune. Oh! the pity of it that every little child cannot grow, live, learn and love among the corn. For the caterpillars never stopped the fun, even the years when they were most numerous.

The eggs laid by my female never hatched, so I do not know this caterpillar in its early stages from experience, but I had enough experience with it in my early stages, that I do not care if I never raise one. No doubt it attains maturity by the same series of moults as the others, and its life history is quite similar. The full-fed caterpillars spin among the leaves on the ground, and with their spines in mind, I would much prefer finding a cocoon, and producing a moth from that stage of its evolution.

The following season I had the good fortune to secure a male and female Io at the same time and by persistence induced them to pose for me on an apple branch. There was no trouble in securing the male as I desired him, with wings folded showing the spots, lining and flushing of colour. But the female was a perverse little body and though I tried patiently and repeatedly she would not lower her wings full width. She climbed around with them three-fourths spread, producing the most beautiful effect of life, but failing to display her striking markings. This is the one disadvantage in photographing moths from life. You secure lifelike effects but sometimes you are forced to sacrifice their wonderful decorations.

CHAPTER IX.
THE SWEETHEART AND THE BRIDE:
CATOCALA AMATYIX—CATOCALA
NEOGAMA

There are no moths so common with us as these, for throughout their season, at any time one is wanted, it is sure to be found either on the sweetbrier clambering over the back wall, among the morning-glories on one side, the wistaria and wild grape on the other, or in the shade of the wild clematis in front. On very sunny days, they leave the shelter of the vines, and rest on the logs of the Cabin close the roof of the verandas. Clinging there they appear like large grey flies, for they are of peculiar shape, and the front wings completely cover the back when in repose. A third or a half of the back wings show as they are lifted to balance the the moths when walking over vines and uncertain footing. They are quite conspicuous on our Cabin, because it is built of the red cedar of Wisconsin; were it of the timber used by our grandfathers, these moths with folded wings would be almost indistinguishable from their surroundings.

Few moths can boast greater beauty. The largest specimen of the 'Sweetheart' that homes with us would measure three and one half inches if it would spread its wings full width as do the moths of other species. No moth is more difficult to describe, because of the delicate blending of so many intangible shades. The front wings are a pale, brownish grey, with irregular markings of tan, and dark splotches outlined with fine deep brown lines. The edges are fluted and escalloped, each raised place being touched with a small spot of tan, and above it a narrow escalloped line of brown. The back wings are bright red, crossed by a circular band of brownish black, three-fourths of an inch from the base, a secondary wider band of the same, and edged with pale yellow.

There is no greater surprise in store for a student of moths than to locate a first Catocala Amatrix, and see the softly blended grey front wings suddenly lift, and the vivid red of the back ones flash out. The under sides of the front wings are a warm creamy tan, crossed by wide bands of dark brown and grey-brown, ending in a delicate grey mist at the edges. The back wings are the same tan shade, with red next the abdomen, and crossed by brown bands of deeper shade than the fore-wings. The shoulders are covered with long silky hair like the front wings. This is so delicate that it becomes detached at the slightest touch of vine or leaf. The abdomen is slightly lighter in colour on top, and a creamy tan beneath. The legs are grey, and the feet to the first joint tan, crossed by faint lines of brown.

The head is small, with big prominent eyes that see better by day than most night moths; for Catocala takes precipitate flight at the merest shadow. The antennae are long, delicate and threadlike, and must be broken very easily in the flight of the moth. It is nothing unusual to see them with one antenna shorter than the other, half, or entirely gone; and a perfect specimen with both antennae, and all the haif on its shoulders, is rare. They have a long tongue that uncoils like Lineata, and Celeus, so they are feeders, but not of day, for they never take flight until evening, except when disturbed. The male is smaller than the female, his fore-wings deeply flushed with darker colour and the back brighter red with more black in the bands.

Neogama, another member of this family, is a degree smaller than Amatrix, but of the same shape. The fore-wings are covered with broken lines of different colours, the groundwork grey, with gold flushings, the lines and dots of the border very like the Sweetheart's. The back wings are pure gold, almost reddish, with dark brownish black bands, and yellow borders. The top of the abdomen is a grey-gold colour. Underneath, the markings are nearly the same as Amatrix, but a gold flush suffuses the moth.

There are numbers of these Catocala moths running the colour scheme of-yellow, from pale chrome to umber. Many shade from light pink through the reds to a dark blood colour. Then there is a smaller number having brown back wings and with others they are white.

The only way I know to photograph them is to focus on some favourable spot, mark the place your plate covers in length and width, and then do your best to coax your subjects in range. If they can be persuaded to walk, they will open their wings to a greater or less degree. A reproduction would do them no sort of justice unless the markings of the back wings show. It is on account of the gorgeous colourings of these that scientists call the species 'afterwings.'

One would suppose that with so many specimens of this beautiful species living with us and swarming the swamp close by, I would be prepared to give their complete life history; but I know less concerning them than any other moths common with us, and all the scientific works I can buy afford little help. Professional lepidopterists dismiss them with few words. One would-be authority disposes of the species with half a dozen lines. You can find at least a hundred Catocala reproduced from museum specimens and their habitat given, in the Holland "Moth Book", but I fail to learn what I most desire to know: what these moths feed on; how late they live; how their eggs appear; where they are deposited; which is their caterpillar; what does it eat; and where and how does it pupate.

Packard, in his "Guide to the Study of Insects", offers in substance this much help upon the subject: "The genus is beautiful, the species numerous, of large size, often three-inch expansion, and in repose form a flat roof. The larva is elongate, slender, flattened beneath and spotted with black, attenuated at each end, with fleshy filaments on the sides above the legs, while the head is flattened and rather forked above. It feeds on trees and rests attached to the trunks. The pupa is covered with a bluish efflorescence, enclosed in a slight cocoon of silk, spun amongst leaves or bark."

This will tend to bear out my contention that scientific works are not the help they should be to the Nature Lover. Heaven save me from starting to locate Catocala moths, eggs, caterpillars or pupae on the strength of this information. I might find moths by accident; nothing on the subject of eggs; neither colour of body, characteristics nor food, to help identify caterpillars; for the statement, 'it feeds on trees,' cannot be considered exactly illuminating when we remember the world full of trees on which caterpillars are feeding; and should one search for cocoon encased pupae among the leaves and bark of tree-tops or earth?

The most reliable information I have had, concerning these moths of which I know least, comes from Professor Rowley. He is the only lepidopterist of four to whom I applied, who could tell me any of the things I am interested in knowing. He writes in substance: "The Bride and Sweetheart are common northern species, as are most of the other members of the group. The Amatrix, with its red wings, is called the Sweetheart because amor means love, and red is love's own colour. The caterpillar feeds on willow. The

Catocala of the yellow "after-wings" is commonly called the Bride, because Neogama, its scientific name, means recently wedded. Its caterpillar feeds on walnut leaves.

"If you will examine the under side of the body of a Catocala moth you will find near the junction of the thorax and abdomen on either side, large open organs reminding one of the ears of a grasshopper, which are on the sides of the first abdominal segment. Examine the bodies of Sphinges and other moths for these same openings. They appear to be ears. Catocala moths feed on juices, and live most of the summer season. Numbers of them have been found sipping sap at a tree freshly cut and you know we take them at night with bait.

"New Orleans sugar and cider or sugar and stale beer are the usual baits. This 'concoction' is put on the bodies of trees with a brush, between eight and ten o'clock at night. During good Catocala years, great numbers of these moths may be taken as they feed at the sweet syrup. So it is proved that their food is sap, honeydew, and other sugary liquids. Mr. George Dodge assures me that he has taken Catocala abbreviatella at milk-weed blooms about eight o'clock of early July evenings. Other species also feed on flowers."

You will observe that in his remarks about the "open organs on the side of the abdominal segment," Professor Rowley may have settled the 'ear' question. I am going to keep sharp watch for these organs, hereafter. I am led to wonder if one could close them in some way and detect any difference in the moth's sense of hearing after having done so.

All of us are enthusiasts about these moths with their modest fore-wings and the gaudy brilliance of the wonderful 'after-wings,' that are so bright as to give common name to the species. We are studying them constantly and hope soon to learn all we care to know of any moths, for our experience with them is quite limited when compared with other visitors from the swamp. But think of the poetry of adding to the long list of birds, animals and insects that temporarily reside with us, a Sweetheart and a Bride!

CHAPTER X.
THE GIANT GAMIN:
TELEA POLYPHEMUS

Time cannot be used to tell of making the acquaintance of this moth until how well worth knowing it is has been explained. That it is a big birdlike fellow, with a six inch sweep of wing, is indicated by the fact that it is named in honour of the giant Polyphemus. Telea means 'the end,' and as scientists fail to explain the appropriateness of this, I am at liberty to indulge a theory of my own. Nature made this handsome moth last, and as it was the end, surpassed herself as a finishing touch on creatures that are, no doubt, her frailest and most exquisite creation.

Polyphemus is rich in shadings of many subdued colours, that so blend and contrast as to give it no superior in the family of short-lived lovers of moonlight. Its front wings are a complicated study of many colours, for some of which it would be difficult to find a name. Really, it is the one moth that must be seen and studied in minutest detail to gain an idea of its beauty. The nearest I can come to the general groundwork of the wing is a rich brown-yellow. The costa is grey, this colour spreading in a widening line from the base of the wing to more than a quarter of an inch at the tip, and closely peppered with black. At the base, the wing is covered with silky yellow-brown hairs. As if to outline the extent of these, comes a line of pinkish white, and then one of rich golden brown, shading into the prevailing colour.

Close the middle of the length of the wing, and half an inch from the costa, is a transparent spot like isinglass, so clear that fine print can be read through it. This spot is outlined with a canary yellow band, and that with a narrow, but sharp circle of black. Then comes a cloudlike rift of golden brown, drifting from the costa across the wing, but, growing fainter until it merges with the general colour near the abdomen. Then half an inch of the yellow-brown colour is peppered with black, similar to the costa; this grows darker until it terminates in a quarter of an inch wide band of almost grey-black crossing the wing. Next this comes a narrower band of pinkish white. The edge begins with a quarter of an inch band of clear yellow-brown, and widens as the wing curves until it is half an inch at the point. It is the lightest colour of rotten apple. The only thing I ever have seen in nature exactly similar was the palest shade of 'mother' found in barrels of vinegar. A very light liver colour comes close it. On the extreme tip is a velvety oval, half black and half pale pink.

The back wings are the merest trifle stronger in this yellow-brown colour, and with the exception of the brown rift are the same in marking, only that all colour, similar to the brown, is a shade deeper.

The 'piece de resistance' of the back wing, is the eyespot. The transparent oval is a little smaller. The canary band is wider, and of stronger colour. The black band around the lower half is yet wider, and of long velvety hairs. It extends in an oval above the transparent spot fully half an inch, then shades through peacock blue, and grey to the hairlike black line enclosing the spot.

The under sides of the wings are pure tan, clouded and lined with shades of rich brown. The transparent spots are outlined with canary, and show a faint line drawn across the middle the long way.

The face is a tiny brown patch with small eyes, for the size of the moth, and large brown antennae, shaped like those of Cecropia. The grey band of the costa crosses the top of the head. The shoulders are covered with pinkish, yellow-brown hair. The top and sides of the abdomen are a lighter shade of the same.

The under side of the abdomen is darker brown, and the legs brown with very dark brown feet. These descriptions do the harmonizing colours of the moth no sort of justice, but are the best I can offer. In some lights it is a rich YELLOW-BROWN, and again a pink flush pervades body and wings.

My first experience with a living Polyphemis (I know Telea is shorter, but it is not suitable, while a giant among moths it is, so that name is best) occurred several years ago. A man brought me a living Polyphemus battered to rags and fringes, antennae broken and three feet missing. He had found a woman trying to beat the clinging creature loose from a door screen, with a towel, before the wings were hardened for flight, and he rescued the remains. There was nothing to say; some people are not happy unless they are killing helpless, harmless creatures; and there was nothing to do.

The moth was useless for a study, while its broken antennae set it crazy, and it shook and trembled continually, going out without depositing any eggs. One thing I did get was complete identification, and another, to attribute the experience to Mrs. Comstock in "A Girl of the Limberlost", when I wished to make her do something particularly disagreeable. In learning a moth I study its eggs, caterpillars, and cocoons, so that fall Raymond and I began searching for Polyphemus. I found our first cocoon hanging by a few threads of silk, from a willow twig overhanging a stream in the limberlost.

A queer little cocoon it was. The body was tan colour, and thickly covered with a white sprinkling like lime. A small thorn tree close the cabin yielded Raymond two more; but these were darker in colour, and each was spun inside three thorn leaves so firmly that it appeared triangular in shape. The winds had blown the cocoons against the limbs and worn away the projecting edges of the leaves, but the midribs and veins showed plainly. In all we had half a dozen of these cocoons gathered from different parts of the swamp, and we found them dangling from a twig of willow or hawthorn, by a small piece of spinning. During the winter these occupied the place of state in the conservatory, and were watched every day. They were kept in the coolest spot, but where the sun reached them at times. Always in watering the flowers, the hose was turned on them, because they would have been in the rain if they had been left out of doors, and conditions should be kept as natural as possible.

Close time for emergence I became very uneasy, because the conservatory was warm; so I moved them to my sleeping room, the coolest in the cabin, where a fireplace, two big windows and an outside door, always open, provide natural atmospheric conditions, and where I would be sure to see them every day. I hung the twigs over a twine stretched from my dresser to the window-sill. One day in May, when the trees were in full bloom, I was working on a tulip bed under an apple tree in the garden, when Molly-Cotton said to me, "How did you get that cocoon in your room wet?"

"I did not water any of the cocoons," I answered. "I have done no sprinkling today. If they are wet, it has come from the inside."

Molly-Cotton dropped her trowel. "One of them was damp on the top before lunch," she cried. "I just now thought of it. The moths are coming!" She started on a run and I followed, but stopped to wash my hands, so she reached them first, and her shout told the news.

"Hurry!" she cried. "Hurry! One is out, and another is just struggling through!" Quickly as I could I stood beside her. One Polyphemus female, a giant indeed, was clinging to a twig with her feet, and from her shoulders depended her wings, wet, and wrinkled as they had been cramped in the pupa case. Even then she had expanded in body until it seemed impossible that she had emerged from the opening of the vacant cocoon. The second one had its front feet and head out, and was struggling frantically to free its shoulders. A fresh wet spot on the top of another cocoon, where the moth had ejected the acid with which it is provided to soften the spinning, was heaving with the pushing head of the third.

Molly-Cotton was in sympathy with the imprisoned moths.

"Why don't you get something sharp, and split the cocoons so they can get out?" she demanded. "Just look at them struggle! They will kill themselves!"

Then I explained to her that if we wanted big, perfect moths we must not touch them. That the evolution of species was complete to the minutest detail. The providence that supplied the acid, required that the moths make the fight necessary to emerge alone, in order to strengthen them so they would be able to walk and cling with their feet, while the wings drooped and dried properly. That if I cut a case, and took out a moth with no effort on its part, it would be too weak to walk, or bear its weight, and so would fall to the floor. Then because of not being in the right position, the wings would harden half spread, or have broken membranes and never develop fully. So instead of doing a kindness I really would work ruination.

"Oh, I see!" cried the wondering girl, and her eyes were large enough to have seen anything, while her brain was racing. If you want to awaken a child and teach it to think, give object lessons such as these, in natural history and study with it, so that every miraculous point is grasped when reached. We left the emerging moths long enough to set up a camera outside, and focus on old tree. Then we hurried back, almost praying that the second moth would be a male, and dry soon enough that the two could be pictured together, before the first one would be strong enough to fly.

The following three hours were spent with them, and every minute enjoyed to the fullest. The first to emerge was dry, and pumping her wings to strengthen them for flight; the second was in condition to pose, but a disappointment, for it was another female. The third was out, and by its smaller size, brighter markings and broad antennae we knew it was a male. His 'antlers' were much wider than those of the first two, and where their markings were pink, his were so vivid as to be almost red, and he was very furry. He had, in fact, almost twice as much long hair as the others, so he undoubtedly was a male, but he was not sufficiently advanced to pose with the females, and I was in doubt as to the wisest course to pursue.

"Hurry him up!" suggested Molly-Cotton. "Tie a string across the window and hang him in the sunshine. I'll bring a fan, and stir the air gently."

This plan seemed feasible, and when the twine was ready, I lifted his twig to place it in the new location. The instant I touched his resting-place and lifted its weight from the twine both females began ejecting a creamy liquid. They ruined the frescoing behind them, as my first Cecropia soiled the lace curtain when I was smaller than Molly-Cotton at that time. We tacked a paper against the wall to prevent further damage. A point to remember in moth culture, is to be ready for this occurrence before they emerge, if you do not want stained frescoing, floors, and hangings.

In the sunshine and fresh air the male began to dry rapidly, and no doubt he understood the presence of his kind, for he was much more active than the females. He climbed the twig, walked the twine body pendent, and was so energetic that we thought we dared not trust him out of doors; but when at every effort to walk or fly he only attempted to reach the females, we concluded that he would not take wing if at liberty. By this time he was fully developed, and so perfect he would serve for a study.

I polished the lenses, focused anew on the tree, marked the limits of exposure, inserted a plate, and had everything ready. Then I brought out the female, Molly-Cotton walking beside me hovering her with a net. The moth climbed from the twig to the tree, and clung there, her wings spread flat, at times setting them quivering in a fluttering motion, or raising them. While Molly-Cotton guarded her I returned for the male, and found him with wings so hardened that could raise them above his back, and lower them full width.

I wanted my study to dignify the term, so I planned it to show the under wings of one moth, the upper of the other. Then the smaller antennae and large abdomen of the female were of interest. I also thought it would be best to secure the male with wings widespread if possible, because his colour was stronger, his markings more pronounced. So I helped the female on a small branch facing the trunk of the tree, and she rested with raised wings as I fervently hoped she would. The male I placed on the trunk, and with wide wings he immediately started toward the female, while she advanced in his direction. This showed his large antennae and all markings and points especially note worthy; being good composition as well, for it centred interest; but there was one objection. It gave the male the conspicuous place and made him appear the larger because of his nearness to the lens and his wing spread; while as a matter of fact, the female had almost an inch more sweep than he, and was bigger at every point save the antennae.

The light was full and strong, the lens the best money could buy, the plate seven by nine inches. By this time long practice had made me rather expert in using my cameras. When the advancing pair were fully inside my circle of focus, I made the first exposure. Then I told Molly-Cotton to keep them as nearly as possible where they were, while I took one breathless peep at the ground glass.

Talk about exciting work! No better focus could be had on them, so I shoved in another plate with all speed, and made a second exposure, which was no better than the first. Had there been time, I would have made a third to be sure, for plates are no object when a study is at all worth while. As a rule each succeeding effort enables you to make some small change for the better, and you must figure on always having enough to lose one through a defective plate or ill luck in development, and yet end with a picture that will serve your purpose.

Then we closed the ventilators and released the moths in the conservatory. The female I placed on a lemon tree in a shady spot, and the male at the extreme far side to see how soon he would find her. We had supposed it would be dark, but they were well acquainted by dusk. The next morning she was dotting eggs over the plants.

The other cocoons produced mostly female living moths, save one that was lost in emergence. I tried to help when it was too late; but cutting open the cocoon afterward proved the moth defective. The wings on one side were only about half size, and on the other little patches no larger than my thumb nail. The body was shrunken and weakly.

At this time, as I remember, Cecropia eggs were the largest I had seen, but these were larger; the same shape and of a white colour with a brown band. The moth dotted them on the under and upper sides of leaves, on sashes and flower pots, tubs and buckets. They turned brown as the days passed. The little caterpillars that emerged from them were reddish brown, and a quarter of an inch long.

I could not see my way to release a small army of two or three hundred of these among my plants, so when they emerged I held a leaf before fifty, that seemed liveliest, and transferred them to a big box. The remainder I placed with less ceremony, over mulberry, elm, maple, wild cherry, grape, rose, apple, and pear, around the Cabin, and gave the ones kept in confinement the same diet.

The leaves given them always were dipped in water to keep them fresh longer, and furnish moisture for the feeders. They grew by a series of moults, like all the others I had raised or seen, and were full size in forty-eight days, but travelled a day or two before beginning the pupa stage of their existence. The caterpillars were big fellows; the segments deeply cut; the bodies yellow-green, with a few sparse scattering hairs, and on the edge of each segment, from a triple row of dots arose a tiny, sharp spine. Each side had series of black touches and the head could be drawn inside the thorax. They were the largest in circumference of any I had raised, but only a little over three inches long.

I arranged both leaves and twigs in the boxes, but they spun among the leaves, and not dangling from twigs, as all the cocoons I had found outdoors were placed previous to that time. Since, I have found them spun lengthwise of twigs in a brush heap. The cocoons of these I had raised were whiter than those of the free caterpillars, and did not have the leaves fastened on the outside, but were woven in a nest of leaves, fastened together by threads.

Polyphemus moths are night flyers, and do not feed. I have tried to tell how beautiful they are, with indifferent success, and they are common with me. Since I learned them, find their cocoons easiest to discover. Through the fall and winter, when riding on trains, I see them dangling from wayside thorn bushes. Once, while taking a walk with Raymond in late November, he located one on a thorn tree in a field beside the road, but he has the eyes of an Indian.

These are the moths that city people can cultivate, for in Indianapolis, in early December, I saw fully one half as many Polyphemus cocoons on the trees as there were Cecropia, and I could have gathered a bushel of them. They have emerged in perfection for me always, with one exception. Personally, I have found more Polyphemus than Cecropia.

These moths are the gamins of their family, and love the streets and lights at night.

Under an arc light at Wabash, Indiana, I once picked up as beautiful a specimen of Polyphemus as I ever saw, and the following day a friend told me that several had been captured the night before in the heart of town.

CHAPTER XI.
THE GARDEN FLY:
PROTOPARCE CAROLINA

Protoparce Carolina is a 'cousin' of Celeus, and so nearly its double that the caterpillars and moths must be seen together to be differentiated by amateurs; while it is doubtful if skilled scientists can always identify the pupa cases with certainty. Carolina is more common in the south, but it is frequent throughout the north. Its caterpillars eat the same food as Celeus, and are the same size. They are a dull green, while Celeus is shining, and during the succession of moults, they show slight variations in colour.

They pupate in a hole in the ground. The moths on close examination show quite a difference from Celeus. They are darker in colour. The fore-wings lack the effect of being laid off in lines. The colour is a mottling of almost black, darkest grey, lighter grey, brown, and white. The back wings are crossed by wavy bands of brownish grey, black, and tan colour, and the yellow markings on the abdomen are larger.

In repose, these moths fold the front wings over the back like large flies. In fact, in the south they are called the 'Tobacco Fly'; and we of the north should add the 'Tomato and Potato Fly.' Because I thought such a picture would be of interest, I reproduced a pair——the male as he clung to a piece of pasteboard in the 'fly' attitude.

Celeus and Carolina caterpillars come the nearest being pests of those of any large moths, because they feed on tomato, potato, and tobacco, but they also eat jimson weed, ground cherry, and several vines that are of no use to average folk.

The Carolina moths come from their pupa cases as featherweights step into the sparring. They feed partially by day, and their big eyes surely see more than those of most other moths, that seem small and deepset in comparison. Their legs are long, and not so hairy as is the rule. They have none of the blind, aimless, helpless appearance of moths that do not feed. They exercise violently in the pupa cases before they burst the shields, and when they emerge their eyes glow and dilate. They step with firmness and assurance, as if they knew where they wanted to go, and how to arrive. They are of direct swift flight, and much experience and dexterity are required to take them on wing.

Both my Carolina moths emerged in late afternoon, about four o'clock, near the time their kind take flight to hunt for food. The light was poor in the Cabin, so I set up my camera and focused on a sweetbrier climbing over the back door.

The newly emerged moth was travelling briskly in that first exercise it takes, while I arranged my camera; so by the time I was ready, it had reached the place to rest quietly until its wings developed. Carolina climbed on my finger with all assurance, walked briskly from it to the roses, and clung there firmly.

The wet wings dropped into position, and the sun dried them rapidly. I fell in love with my subject. He stepped around so jauntily in comparison with most moths. The picture he made while clinging to the roses during the first exposure was lovely.

His slender, trim legs seemed to have three long joints, and two short in the feet. In his sidewise position toward the lens, the abdomen showed silver-white beneath, silvery grey on the sides, and large patches of orange surrounded by black, with touches of white on top. His wings were folded together on his back as they drooped, showing only the under sides, and on these the markings were more clearly defined than on top. In the sunlight the fore pair were a warm tan grey, exquisitely lined and shaded. They were a little more than half covered by the back pair, that folded over them. These were a darker grey, with tan and almost black shadings, and crossed by sharply zig-zagging lines of black. The grey legs were banded by lines of white. The first pair clung to the stamens of the rose, the second to the petals, and the third stretched out and rested on a leaf.

There were beautiful markings of very dark colour and white on the thorax, head, shoulders, and back wings next the body. The big eyes, quite the largest of any moth I remember, reminded me of owl eyes in the light. The antennae, dark, grey-brown on top, and white on the under side, turned back and drooped beside the costa, no doubt in the position they occupied in the pupa case.

The location was so warm, and the moth dried so rapidly, that by the time two good studies were made of him in this position, he felt able to step to some leaves, and with no warning whatever, reversed his wings to the 'fly' position, so that only the top side of the front pair showed. The colour was very rich and beautiful, but so broken in small patches and lines, as to be difficult to describe. With the reversal of the wings the antennae flared a little higher, and the exercise of the sucking tube began. The moth would expose the whole length of the tube in a coil, which it would make larger and contract by turns, at times drawing it from sight. When it was uncoiled the farthest, a cleft in the face where it fitted could be seen.

The next day my second Carolina case produced a beautiful female. The history of her emergence was exactly similar to that of the male. Her head, shoulders, and abdomen seemed nearly twice the size of his, while her wings but a trifle, if any larger.

As these moths are feeders, and live for weeks, I presume when the female has deposited her eggs, the abdomen contracts, and loses its weight so that she does not require the large wings of the females that only deposit their eggs and die. They are very heavy, and if forced to flight must have big wings to support them. I was so interested in this that I slightly chloroformed the female, and made a study of the pair. The male was fully alive and alert, but they had not mated, and he would not take wing. He clung in his natural position, so that he resembled a big fly, on the smooth side of the sheet of corrugated paper on which I placed the female. His wings folded over each other. The abdomen and the antennae were invisible, because they were laid flat on the costa of each wing.

The female clung to the board, in any position in which she was placed. Her tongue readily uncoiled, showing its extreme length, and curled around a pin. With a camel'shair brush I gently spread her wings to show how near they were the size of the male's, and how much larger her body was.

Her fore-wings were a trifle lighter in colour than the male's, and not so broken with small markings. The back wings were very similar. Her antennae stood straight out from the head on each side, of their own volition and differed from the male's. It has been my observation that in repose these moths fold the antennae as shown by the male. The position of the female was unnatural. In flight, or when feeding, the antennae are raised, and used as a guide in finding food flowers. A moth with broken antennae seems dazed and helpless, and in great distress.

151

I have learned by experience in handling moths, that when I induce one to climb upon bark, branch, or flower for a study, they seldom place their wings as I want them. Often it takes long and patient coaxing, and they are sensitive to touch. If I try to force a fore-wing with my fingers to secure a wider sweep, so that the markings of the back wings show, the moths resent it by closing them closer than before, climbing to a different location or often taking flight.

But if I use a fine camel's-hair brush, that lacks the pulsation of circulation, and gently stroke the wing, and sides of the abdomen, the moths seems to like the sensation and grow sleepy or hypnotized. By using the brush I never fail to get wing extension that will show markings, and at the same time the feet and body are in a natural position. After all is said there is to say, and done there is to do, the final summing up and judgment of any work on Natural History will depend upon whether it is true to nature. It is for this reason I often have waited for days and searched over untold miles to find the right location, even the exact leaf, twig or branch on which a subject should be placed.

I plead guilty to the use of an anesthetic in this chapter only to show the tongue extension of Carolina, because it is the extremest with which I am acquainted; and to coaxing wide wing sweep with the camel'shair brush; otherwise either the fact that my subjects are too close emergence ever to have taken flight, or sex attraction alone holds them.

If you do not discover love running through every line of this text and see it shining from the face of each study and painting, you do not read aright and your eyes need attention. Again and again to the protests of my family, I have made answer—

"To work we love we rise betimes, and go to it with delight."

From the middle of May to the end of June of the year I was most occupied with this book, my room was filled with cocoons and pupa cases. The encased moths I had reason to believe were on the point of appearing lay on a chair beside my bed or a tray close my pillow. That month I did not average two hours of sleep in a night, and had less in the daytime. I not only arose 'betimes,' but at any time I heard a scratching and tugging moth working to enter the world, and when its head was out, I was up and ready with note-book and camera. Day helped the matter but slightly, for any moth emerging in the night had to be provided a location, and pictured before ten o'clock or it was not safe to take it outside. Then I had literally 'to fly' to develop the plate, make my print and secure exact colour reproduction while the moth was fresh.

For this is a point to remember in photographing a moth. A FREE LIVING MOTH NEVER RAISES ITS WINGS HIGHER THAN A STRAIGHT LINE FROM THE BASES CROSSING THE TOP OF THE THORAX.

It requires expert and adept coaxing to get them horizontal with their bases. If you do, you show all markings required; and preserve natural values, quite the most important things to be considered.

I made a discovery with Carolina. Moths having digestive organs and that are feeders are susceptible to anaesthetics in a far higher degree than those that do not feed. Many scientific workers confess to having poured full strength chloroform directly on nonfeeders, mounted them as pinned specimens and later found them living; so that sensitive lepidopterists have abandoned its use for the cyanide or gasoline jar. I intended to give only a whiff of chloroform to this moth, just enough that she would allow her tongue to remain uncoiled until I could snap its fullest extent, but I could not revive her. The same amount would have had no effect whatever on a non-feeder.

CHAPTER XII.
BLOODY-NOSE OF SUNSHINE HILL: HEMARIS THYSBE

John Brown lives a mile north of our village, in the little hamlet of Ceylon. Like his illustrious predecessor of the same name he is willing to do something for other people. Mr. Brown owns a large farm, that for a long distance borders the Wabash River where it is at its best, and always the cameras and I have the freedom of his premises.

On the east side of the village, about half its length, swings a big gate, that opens into a long country lane. It leads between fields of wheat and corn to a stretch of woods pasture, lying on a hillside, that ends at the river. This covers many acres, most of the trees have been cut; the land rises gradually to a crest, that is crowned by a straggling old snake fence, velvety black in places, grey with lint in others, and liberally decorated its entire length with lichens, in every shade of grey and green. Its corners are filled with wild flowers, ferns, gooseberries, raspberries, black and red haw, papaw, wild grapevines, and trees of all varieties. Across the fence a sumac covered embankment falls precipitately to the Wabash, where it sweeps around a great curve at Horseshoe Bend. The bed is stone and gravel, the water flows shallow and pure in the sunlight, and mallows and willows fringe the banks.

Beside this stretch of river most of one summer was spent, because there were two broods of cardinals, whose acquaintance I was cultivating, raised in those sumacs. The place was very secluded, as the water was not deep enough for fishing or swimming. On days when the cardinals were contrary, or to do the birds justice, when they had experiences with an owl the previous night, or with a hawk in the morning, and were restless or unduly excited, much grist for my camera could be found on the river banks.

These were the most beautiful anywhere in my locality. The hum of busy life was incessant. From the top twig of the giant sycamore in Rainbow Bottom, the father of the cardinal flock hourly challenged all creation to contest his right to one particular sumac. The cardinals were the attraction there; across the fence where the hill sloped the length of the pasture to the lane, lures were many and imperative. Despite a few large trees, compelling right to life by their majesty, that hillside was open pasture, where the sunshine streamed all day long. Wild roses clambered over stumps of fallen monarchs, and scrub oak sheltered resting sheep. As it swept to the crest, the hillside was thickly dotted with mullein, its pale yellow-green leaves spreading over the grass, and its spiral of canary-coloured bloom stiffly upstanding.

There were thistles, the big, rank, richly growing, kind, that browsing cattle and sheep circled widely.

Very beautiful were these frosted thistles, with their large, widespreading base leaves, each spine needle-tipped, their uplifted heads of delicate purple bloom, and their floating globes of silken down, with a seed in their hearts. No wonder artists have painted them, decorators conventionalized them; even potters could not pass by their artistic merit, for I remembered that in a china closet at home there were Belleck cups moulded in the shape of a thistle head.

Experience had taught me how the appreciate this plant. There was a chewink in the Stanley woods, that brought off a brood of four, under the safe shelter of a rank thistle leaf, in the midst of trampling herds of cattle driven wild by flies. There was a ground sparrow near the Hale sand pit, covered by a base leaf of another thistle, and beneath a third on Bob's lease, I had made a study of an exquisite nest. Protection from the rank leaves was not all the birds sought of these plants, for goldfinches were darting around inviting all creation to "See me?" as they gathered the silken down for nest lining. Over the sweetly perfumed purple heads, the humming-birds held high carnival on Sunshine Hillside all the day. The honey and bumble bees fled at the birds' approach, but what were these others, numerous everywhere, that clung to the blooms, greedily thrusting their red noses between the petals, and giving place to nothing else?

For days as I passed among them, I thought them huge bees. The bright colouring of their golden olive-green, and red-wine striped bodies had attracted me in passing. Then one of them approached a thistle head opposite me in such a way its antennae and the long tongue it thrust into the bloom could be seen. That proved it was not a bee, and punishment did not await any one who touched it.

There were so many that with one sweep of the net two were captured. They were examined to my satisfaction and astonishment. They were moths! Truly moths, feeding in the brilliant sunshine all the day; bearing a degree of light and heat I never had known any other moth to endure. Talk about exquisite creatures! These little day moths, not much larger than the largest bumble bees, had some of their gaudiest competitors of moonlight and darkness outdone.

The head was small and pointed, with big eyes, a long tongue, clubbed antennae, and a blood-red nose. The thorax above was covered with long, silky, olive-green hair; the top of the abdomen had half an inch band of warm tan colour, then a quarter of an inch band of velvety red wine, then a band nearer the olive of the shoulders. The males had claspers covered with small red-wine feathers tan tipped. The thorax was cream-coloured below and the under side of the abdomen red wine crossed with cream-coloured lines at each segment.

The front wings had the usual long, silky hairs. They were of olive-green shading into red, at the base, the costa was red, and an escalloped band of red bordered them. The intervening space was transparent like thinnest isinglass, and crossed with fine red veins. The back wings were the same, only the hairs at the base were lighter red, and the band at the edge deeper in colour.

The head of the male seemed sharper, the shoulders stronger olive, the wings more pointed at the apex, where the female's were a little rounded. The top of the abdomen had the middle band of such strong red that it threw the same colour over the bands above and below it; giving to the whole moth a strong red appearance when on wing. They, were so fascinating the birds were forgotten, and the hillside hunted for them until a pair were secured to carry home for identification, before the whistle of the cardinal from Rainbow Bottom rang so sharply that I remembered this was the day I had hoped to secure his likeness; and here I was allowing a little red-nosed moth so to thrust itself upon my attention, that my cameras were not even set up and focused on the sumac.

This tiny sunshine moth, Hemaris Thysbe, was easy of identification, and its whole life history before me on the hillside. I was too busy with the birds to raise many caterpillars, so reference to several books taught me that they all agreed on the main points of Hemaris history.

Hemaris means 'bloody nose.' 'Bloody nose' on account of the red first noticed on the face, though some writers called them 'Clear wings,' because of the transparent spaces on the wings. Certainly 'clear wings' is a most appropriate and poetic name for this moth. Fastidious people will undoubtedly prefer it for common usage. For myself, I always think of the delicate, gaudy little creature, greedily thrusting its blood-red nose into the purple thistle blooms; so to my thought it returns as 'bloody nose.'

The pairs mate early after emerging, and lay about two hundred small eggs to the female, from which the caterpillars soon hatch, and begin their succession of moults. One writer gave black haw and snowball as their favourite foods, and the length of the caterpillar when full grown nearly two inches. They are either a light brown with yellow markings, or green with yellow; all of them have white granules on the body, and a blue-black horn with a yellow base. They spin among the leaves on the ground, and the pupa, while small, is shaped like Regalis, except that it has a sharper point at each end, and more prominent wing shields. It has no raised tongue case, although it belongs to the family of 'long tongues.'

On learning all I could acquire by experience with these moths, and what the books had to teach, I became their warm admirer. One sunny morning climbing the hill on the way to the cardinals, with fresh plates in my cameras, and high hopes in my heart, I passed an unsually large fine thistle, with half a dozen Thysbe moths fluttering over it as if nearly crazed with fragrance, or honey they were sipping.

"Come here! Come here! Come here!" intoned the cardinal, from the sycamore of Rainbow Bottom.

"Just you wait a second, old fellow!" I heard myself answering. Scarcely realizing what I was doing, the tripod was set up, the best camera taken out, and focused on that thistle head. The moths paid no attention to bees, butterflies, or humming-birds visiting the thistle, but this was too formidable, and by the time the choicest heads were in focus, all the little red fellows had darted to another plant. If the camera was moved there, they would change again, so I sat in the shade of a clump of papaws to wait and see if they would not grow accustomed to it.

They kept me longer than I had expected, and the chances are I would have answered the cardinal's call, and gone to the river, had it not been for the interest found in watching a beautiful grey squirrel that homed in an ivy-covered stump in the pasture. He seemed to have much business on the fence at the hilltop, and raced back and forth to it repeatedly. He carried something, I could not always tell what, but at times it was green haws. Once he came with no food, and at such a headlong run that he almost turned somersaults as he scampered up the tree.

For a long time he was quiet, then he cautiously peeped out. After a while he ventured to the ground, raced to a dead stump, and sitting on it, barked and scolded with all his might. Then he darted home again. When he had repeated this performance several times, the idea became apparent. There was some danger to be defied in Rainbow Bottom, but not a sound must be made from his home. The bark of a dog hurried me to the fence in time to see some hunters passing in the bottom, but I thanked mercy they were on the opposite side of the river and it was not probable they would wade, so my birds would not be disturbed. When the squirrel felt that he must bark and chatter, or burst with tense emotions, he discreetly left his mate and nest. I did some serious thinking on the 'instinct' question. He might choose a hollow log for his home by instinct, or eat certain foods because hunger urged him, but could instinct teach him not to make a sound where his young family lay? Without a doubt, for this same reason, the cardinal sang from every tree and bush around Horseshoe Bend, save the sumac where his mate hovered their young.

The matter presented itself in this way. The squirrel has feet, and he runs with them. He has teeth, and he eats with them. He has lungs, and he breathes with them. Every organ of his interior has its purpose, and is used to fulfil it. His big, prominent eyes come from long residence in dark hollows. His bushy tail helps him in long jumps from tree to tree.

Every part of his anatomy is created, designed and used to serve some purpose, save only his brain, the most complex and complicated part of him. Its only use and purpose is to form one small 'tidbit' for the palate of the epicure! Like Sir Francis, who preached a sermon to the birds, I found me delivering myself of a lecture to the squirrels, birds, and moths of Sunshine Hill. The final summing up was, that the squirrel used his feet, teeth, eyes and tail; that could be seen easily, and by his actions it could be seen just as clearly that he used his brain also.

There was not a Thysbe in front of the lens, so picking up a long cudgel I always carry afield, and going quietly to surrounding thistles, I jarred them lightly with it, and began rounding up the Hemaris family in the direction of the camera. The trick was a complete success. Soon I had an exposure on two. After they had faced the camera once, and experienced no injury, like the birds, they accepted it as part of the landscape. The work was so fascinating, and the pictures on the ground glass so worth while, that before I realized what I was doing, half a dozen large plates were gone, and for this reason, work with the cardinals that day ended at noon. This is why I feel that at times in bird work the moths literally 'thrust themselves' upon me.

CHAPTER XIII.
THE MODEST MOTH:
TRIPTOGON MODESTA

Of course this moth was named Modesta because of modest colouring. It reminds me of a dove, being one of my prime favourites. On wing it is suggestive of Polyphemus, but its colours are lighter and softer. Great beauty that Polyphemus is, Modesta equals it.

Modesta belongs to the genus Triptogon, species Modesta—hence the common name, the Modest moth. I am told that in the east this moth is of stronger colouring than in the central and western states. I do not know about the centre and west, but I do know that only as far east as Indiana, Modesta is of more delicate colouring than it is described by scientists of New York and Pennsylvania; and, of course, as in almost every case, the female is not so strongly coloured as the male.

I can class the Modest moth and its caterpillar among those I know, but my acquaintance with it is more limited than with almost any other. My first introduction came when I found a caterpillar of striking appearance on water sprouts growing around a poplar stump in a stretch of trees beside the Wabash. I carried it home with a supply of the leaves for diet, but as a matter of luck, it had finished eating, and was ready to pupate. I write of this as good luck, because the poplar tree is almost extinct in my location. I know of only one in the fields, those beside the river, and a few used for ornamental shade trees. They are so scarce I would have had trouble to provide the caterpillar with natural food; so I was glad that it was ready to pupate when found.

Any one can identify this caterpillar easily, as it is most peculiar. There is a purplish pink cast on the head and mouth of the full-grown caterpillar, and purplish red around the props. The body is a very light blue-green, faintly tinged with white, and yellow in places. On the sides are white obliques, or white, shaded with pink, and at the base of these, a small oval marking. There is a small short horn on the head. But the distinguishing mark is a mass of little white granules, scattered all over the caterpillar. It is so peppered with these, that failure to identify it is impossible.

These caterpillars pupate in the ground. I knew that, but this was before I had learned that the caterpillar worked out a hole in the ground, and the pupa case only touched the earth upon which it lay. So when my Modesta caterpillar ceased crawling, lay quietly, turned dark, shrank one half in length, and finally burst the dead skin, and emerged in a shining dark brown pupa case two inches long, I got in my work. I did well. A spade full of garden soil was thoroughly sifted, baked in the oven to kill parasites and insects, cooled, and put in a box, and the pupa case buried in it. Every time it rained, I opened the box, and moistened the earth. Two months after time for emergence, I dug out the pupa case to find it white with mould. I had no idea what the trouble was, for I had done much work over that case, and the whole winter tended it solicitously. It was one of my earliest attempts, and I never have found another caterpillar, or any eggs, though I often search the poplars for them.

However, something better happened. I say better, because I think if they will make honest confession, all people who have gathered eggs and raised caterpillars from them in confinement, by feeding cut leaves, will admit that the pupa cases they get, and the moths they produce are only about half size. The big fine cases and cocoons are the ones you find made by caterpillars in freedom, or by those that have passed at least the fourth or fifth moult out of doors. So it was a better thing for my illustration, and for my painting, when in June of this year, Raymond, in crossing town from a ball game, found a large, perfect Modesta female. He secured her in his hat, and hurried to me. Raymond's hat has had many wonderful things in it besides his head, and his pockets are always lumpy with boxes.

Although perfect, she had mated, deposited her eggs, and was declining. All she wanted was to be left alone, and she would sit with wings widespread wherever placed. I was in the orchard, treating myself to some rare big musky red raspberries that are my especial property, when Raymond came with her. He set her on a shoot before me, and guarded her while I arranged a camera. She was the most complacent subject I ever handled outdoors, and did not make even an attempt to fly. Raymond was supposed to be watching while I worked, but our confidence in her was so great, that I paid all my attention to polishing my lenses, and getting good light, while Raymond gathered berries with one hand, and promiscuously waved the net over the bushes with the other.

During the first exposure, Modesta was allowed to place and poise herself as seemed natural. For a second, I used the brush on her gently, and coaxed her wings into spreading a little wider than was natural. These positions gave every evidence of being pleasing and yet I was not satisfied. There was something else in the back of my head that kept obtruding itself as I walked to the Cabin, with the beautiful moth clinging to my fingers. I did not feel quite happy about her, so she was placed in a large box, lined with corrugated paper, to wait a while until the mist in my brain cleared, and my nebulous disturbance evolved an idea. It came slowly. I had a caterpillar long ago, and had investigated the history of this moth. I asked Raymond where he found her and he said, "Coming from the game." Now I questioned him about the kind of a tree, and he promptly answered, "On one of those poplars behind the schoolhouse."

That was the clue. Instantly I recognized it. A poplar limb was what I wanted. Its fine, glossy leaf, flattened stem, and smooth upright twigs made a setting, appropriate, above all others, for the Modest moth.

I explained the situation to the Deacon, and he had Brenner drive with him to the Hirschy farm, and help secure a limb from one of the very few Lombardy poplars of this region. They drove very fast, and I had to trouble to induce Modesta to clamber over a poplar twig, and settle. Then by gently stroking, an unusual wing sweep was secured, because there is a wonderful purple-pink and a peculiar blue on the back wings.

It has been my experience that the longer a moth of these big short-lived subjects remains out of doors, the paler its colours become, and most of them fade rapidly when mounted, if not kept in the dark. So my Modesta may have been slightly faded, but she could have been several shades paler and yet appeared most beautiful to me.

Her head, shoulders, and abdomen were a lovely dove grey; that soft tan grey, with a warm shade, almost suggestive of pink. I suppose the reason I thought of this was because at the time two pairs of doves, one on a heap of driftwood overhanging the river, and the other in an apple tree in the Aspy orchard a few rods away, were giving me much trouble, and I had dove grey on my mind.

This same dove grey coloured the basic third of the fore-wings. Then they were crossed with a band only a little less in width, of rich cinnamon brown. There was a narrow wavy line of lighter brown, and the remaining third of the wing was paler, but with darker shadings. These four distinct colour divisions were exquisitely blended, and on the darkest band, near the costa, was a tiny white half moon. The under sides of the fore-wings were a delicate brownish grey, with heavy flushings of a purplish pink, a most beautiful colour.

The back wings were dove colour near the abdomen, more of a mouse colour around the edges, and beginning strongly at the base, and spreading in lighter shade over the wing, was the same purplish pink of the front under-wing, only much stronger. Near the abdomen, a little below half the length, and adjoining the grey; each wing had a mark difficult to describe in shape, and of rich blue colour.

The antennae stood up stoutly, and were of dove grey on one side, and white on the other. The thorax, legs, and under side of the abdomen were more of the mouse grey in colour. Over the whole moth in strong light, there was an almost intangible flushing of palest purplish pink. It may have shaded through the fore-wing from beneath, and over the back wing from above. At any rate, it was there, and so lovely and delicate was the whole colour scheme, it made me feel that I would give much to see a newly emerged male of this species. In my childhood my mother called this colour aniline red.

I once asked a Chicago importer if he believed that Oriental rug weavers sometimes use these big night moths as colour guides in their weaving. He said he had heard this, and gave me the freedom of his rarest rugs. Of course the designs woven into these rugs have a history, and a meaning for those who understand. There were three, almost priceless, one of which I am quite sure copied its greys, terra cotta, and black shades from Cecropia.

There was another, a rug of pure silk, that never could have touched a floor, or been trusted outside a case, had it been my property, that beyond all question took its exquisite combinations of browns and tans with pink lines, and peacock blue designs from Polyphemus. A third could have been copied from no moth save Modesta, for it was dove grey, mouse grey, and cinnamon brown, with the purplish pink of the back wings, and exactly the blue of their decorations. Had this rug been woven of silk, as the brown one, that moment would have taught me why people sometimes steal when they cannot afford to buy. Examination of the stock of any importer of high grade rugs will convince one who knows moths, that many of our commonest or their near relatives native to the Orient are really used as models for colour combinations in rug weaving. The Herat frequently has moths in its border.

The Modest moth has a wing sweep in large females of from five and one-half to six inches. In my territory they are very rare, only a few caterpillars and one moth have fallen to me. This can be accounted for by the fact that the favourite food tree of the caterpillar is so scarce, for some reason having become almost extinct, except in a few cases where they are used for shade.

The eggs are a greyish green, and have the pearly appearance of almost all moth eggs. On account of white granules, the caterpillar cannot fail to be identified. The moths in their beautiful soft colouring are well worth search and study. They are as exquisitely shaded as any, and of a richness difficult to describe.

CHAPTER XIV.
THE PRIDE OF THE LILACS:
ATTACUS PROMETHEA

So far as the arrangement of the subjects of this book in family groupings is concerned, any chapter might come first or last. It is frankly announced as the book of the Nature Lover, and as such is put together in the form that appears to me easiest to comprehend and most satisfying to examine. I decided that it would be sufficient to explain the whole situation to the satisfaction of any one, if I began the book with a detailed history of moth, egg, caterpillar, and cocoon and then gave complete portrayal of each stage in the evolution of one cocoon and one pupa case moth. I began with Cecropia, the commonest of all and one of the most beautiful for the spinners, and ended with Regalis, of earth—and the rarest.

The luck I had in securing Regalis in such complete form seems to me the greatest that ever happened to any, worker in this field, and it reads more like a fairy tale than sober every-day fact, copiously illustrated with studies from life. At its finish I said, "Now I am done. This book is completed." Soon afterward, Raymond walked in with a bunch of lilac twigs in his hand from which depended three rolled leaves securely bound to their twigs by silk spinning.

"I don't remember that we ever found any like these," he said. 'Would you be interested in them?'

Would I? Instantly I knew this book was not finished. As I held the firm, heavy, leaf-rolled cocoons in my hand, I could see the last chapter sliding over from fourteen to fifteen to make place for Promethea, the loveliest of the Attacine group, a cousin of Cecropia. Often I had seen the pictured cocoon, in its neat little, tight little leaf-covered shelter, and the mounted moths of scientific collections and museums; I knew their beautiful forms and remembered the reddish tinge flushing the almost black coat of the male and the red wine and clay-coloured female with her elaborate marks, spots, and lines. Right there the book stopped at leaf-fall early in November to await the outcome of those three cocoons. If they would yield a pair in the spring, and if that pair would emerge close enough together to mate and produce fertile eggs, then by fall of the coming year I would have a complete life history. That was a long wait, thickly punctuated with 'ifs.'

Then the twig was carried to my room and stood in a vase of intricate workmanship and rare colouring.

Every few days I examined those cocoons and tested them by weight. I was sure they were perfect. That spring I had been working all day and often at night, so I welcomed an opportunity to spend a few days at a lake where I would meet many friends; boating and fishing were fine, while the surrounding country was one uninterrupted panorama of exquisite land and water pictures. I packed and started so hastily I forgot my precious cocoons. Two weeks later on my return, before I entered the Cabin, I walked round it to see if my flowers had been properly watered and tended. It was not later than three in the afternoon but I saw at least a dozen wonderful big moths, dusky and luring, fluttering eagerly over the wild roses covering a south window of the Deacon's room adjoining mine on the west. Instantly I knew what that meant. I hurried to the room and found a female Promothea at the top of the screen covering a window that the caretaker had slightly lowered. I caught up a net and ran to bring a step-ladder. The back foundation is several feet high and that threw the tops of the windows close under the eaves. I mounted to the last step and balancing made a sweep to capture a moth. They could see me and scattered in all directions. I waited until they were beginning to return, when from the thicket of leaves emerged a deep rose-flushed little moth that sailed away, with every black one in pursuit. I almost fell from the ladder. I went inside, only to learn that what I feared was true. The wind had loosened the screen in my absence, and the moth had passed through a crack, so narrow it seemed impossible for it to escape.

Only those interested as I was, and who have had similar experience, know how to sympathize. I had thought a crowbar would be required to open one of those screens!

With sinking heart I hurried to my room. Joy! There was yet hope! The escaped moth was the only one that had emerged. The first thing was to fasten the screen, the next to live with the remaining cocoons.

The following morning another, female appeared, and a little later a male.

The cocoons were long, slender, closely leaf-wrapped and hung from stout spinning longer than the average leaf stem. The outside leaf covering easily could be peeled away as the spinning did not seem to adhere except at the edges. There was a thin waterproof coating as with Cecropia, then a little loose spinning that showed most at top and bottom, the leaf wrapping being so closely drawn that it was plastered against the body of the heavy inner case around the middle until it adhered. The inner case was smooth and dark inside and the broken pupa case nearly black.

The male and female differed more widely in colour and markings than any moths with which I had worked. At a glance, the male reminded me of a monster Mourning Cloak butterfly. The front wings from the base extending over half the surface were a dark brownish black, outlined with a narrow escalloped line of clay colour of light shade. The black colour from here lightened as it neared the margin. At the apex it changed to a reddish brown tinge that surrounded the typical eye-spot of all the Attacus group for almost three-fourths of its circumference. The bottom of the eye was blackish blue, shading abruptly to pale blue at the top. The straggle M of white was in its place at the extreme tip, on the usual rose madder field. From there a broad clay-coloured band edged the wing and joined the dark colour in escallops. Through the middle of it in an irregular wavy line was traced an almost hair-fine marking of strong brown. The back wings were darker than the darkest part of the fore-wings and this colour covered them to the margin, lightening very slightly. A clay-coloured band bordered the edge, touched with irregular splashes of dark brown, a little below them a slightly heavier line than that on the fore-wing, which seemed to follow the outline of the decorations.

Underneath, the wings were exquisitely marked, flushed, and shaded almost past description in delicate and nearly intangible reddish browns, rose madder on grey, pink-tinged brown and clay colour. On the fore-wings the field from base to first line was reddish brown with a faint tinge of tan beside the costa. From this to the clay-coloured border my descriptive powers fail. You could see almost any shade for which you looked. There were greyish places flushed with scales of red and white so closely set that the result was frosty pink. Then the background would change to brown with the same over-decoration. The bottom of the eye-spot was dark only about one-fourth the way, the remaining three-fourths, tan colour outlined at the top with pale blue and black in fine lines. The white M showed through on a reddish background, as did the brown line of the clay border. The back wings widespread were even lovelier. Beginning about the eighth of an inch from the top was a whitish line tracing a marking that when taken as a whole on both outspread wings, on some, slightly resembled a sugar maple leaf, and on others, the perfect profile of a face. There was a small oblong figure of pinkish white where the eye would fall, and the field of each space was brownish red velvet. From this to the clay-coloured band with its paler brown markings and lines, the pink and white scales sprinkled the brown ground; most of the pink, around the marking, more of the white, in the middle of the space; so few of either, that it appeared to be brown where the clay border joined.

The antennae were shaped as all of the Attacus group, but larger in proportion to size, for my biggest Promethea measured only four and a quarter from tip to tip, and for his inches carried larger antlers than any Cecropia I ever saw of this measurement, those of the male being very much larger than the female. In colour they were similar to the darkest part of the wings, as were the back of the head, thorax and abdomen. The hair on the back of the thorax was very long. The face wore a pink flush over brown, the eyes bright brown, the under thorax covered with long pinkish brown hairs, and the legs the same. A white stripe ran down each side of the abdomen, touched with a dot of brownish red wine colour on the rings. The under part was pinkish wine crossed with a narrow white line at each segment. The claspers were prominent and sharp. The finishing touch of the exquisite creation lay in the fact that in motion, in strong light the red wine shadings of the under side cast an intangible, elusive, rosy flush over the dark back of the moth that was the mast delicate and loveliest colour effect I ever have seen on marking of flower, bird, or animal.

For the first time in all my experience with moths the female was less than the male.

Even the eggs of this mated pair carried a pinkish white shade and were stained with brown. They were ovoid in shape and dotted the screen door in rows. The tiny caterpillars were out eleven days later and proved to be of the kind that march independently from their shells without stopping to feed on them. Of every food offered, the youngsters seemed to prefer lilac leaves; I remembered that they had passed the winter wrapped in these, dangling from their twigs, and that the under wings of the male and much of the female bore a flushing of colour that was lilac, for what else is red wine veiled with white? So I promptly christened them, 'The Pride of the Lilacs.' They were said to eat ash, apple pear, willow, plum, cherry, poplar and many other leaves, but mine liked lilac, and there was a supply in reach of the door, so they undoubtedly were lilac caterpillars, for they had nothing else to eat.

The little fellows were pronouncedly yellow. The black head with a grey stripe joined the thorax with a yellow band. The body was yellow with black rings, the anal parts black, the legs pale greyish yellow. They made their first moult on the tenth day and when ready to eat again they were stronger yellow than before, with many touches of black. They moulted four times, each producing slight changes until the third, when the body took on a greenish tinge, delicate and frosty in appearance. The heads were yellow with touches of black, and the anal shield even stronger yellow, with black. At the last moult there came a touch of red on the thorax, and of deep blue on the latter part of the body.

In spinning they gummed over the upper surface of a leaf and, covering it with silk, drew it together so that nothing could be seen of the work inside. They began spinning some on the forty-second, some on the forty-third day, when about three inches in length and plump to bursting. I think at a puncture in the skin they would have spurted like a fountain. They began spinning at night and were from sight before I went to them the following morning. So I hunted a box and packed them away with utmost care.

I selected a box in which some mounted moths had been sent me by a friend in Louisiana, and when I went to examine my cocoons toward spring, to my horror I found the contents of the box chopped to pieces and totally destroyed. Pestiferous little 'clothes' moths must have infested the box, for there were none elsewhere in the Cabin. For a while this appeared to be too bad luck; but when luck turns squarely against you, that is the time to test the essence and quality of the word 'friend.' So I sat me down and wrote to my friend, Professor Rowley, of Missouri, and told him I wanted Promethea for the completion of this book; that I had an opportunity to make studies of them and my plate was light-struck, and house-moths had eaten my cocoons. Could he do anything? To be sure he could. I am very certain he sent me two dozen 'perfectly good' cocoons.

From the abundance of males that have come to seek females of this species at the Cabin, ample proof seems furnished that they are a very common Limberlost product; but I never have found, even when searching for them, or had brought to me a cocoon of this variety, save the three on one little branch found by Raymond, when he did not know what they were. Because of the length of spinning which these caterpillars use to attach their cocoons, they dangle freely in the wind, and this gives them especial freedom from attack.

CHAPTER XV.
THE KING OF THE POETS:
CITHERONIA REGALIS

To the impetuosity of youth I owe my first acquaintance with the rarest moth of the Limberlost; "not common anywhere," say scientific authorities. Molly-Cotton and I were driving to Portland-town, ten miles south of our home. As customary, I was watching fields, woods, fence corners and roadside in search of subjects; for many beautiful cocoons and caterpillars, much to be desired, have been located while driving over the country on business or pleasure.

With the magnificent independence of the young, Molly-Cotton would have scouted the idea that she was searching for moths also, but I smiled inwardly as I noticed her check the horse several times and scan a wayside bush, or stretch of snake fence. We were approaching the limits of town, and had found nothing; a slow rain was falling, and the shimmer on bushes and fences made it difficult to see objects plainly. Several times I had asked her to stop the horse, or drive close the fields when I was sure of a moth or caterpillar, though it was very late, being close the end of August; but we found only a dry leaf, or some combination that had deceived me.

Just on the outskirts of Portland, beside a grassy ditch and at the edge of a cornfield, grew a cluster of wild tiger lilies. The water in the ditch had kept them in flower long past their bloomtime. On one of the stems there seemed to be a movement.

"Wait a minute!" I cried, and Molly-Cotton checked the horse, but did not stop, while I leaned forward and scanned the lilies carefully. What I thought I saw move appeared to be a dry lily bloom of an orange-red colour, that had fallen and lodged on the grasses against a stalk.

"It's only a dead lily," I said; "drive on."

"Is there a moth that colour?" asked Molly-Cotton.

"Yes," I replied. "There is an orange-brown species, but it is rare. I never have seen a living one."

So we passed the lilies. A very peculiar thing is that when one grows intensely interested in a subject, and works over it, a sort of instinct, an extra sense as it were, is acquired. Three rods away, I became certain I had seen something move, so strongly the conviction swept over me that we had passed a moth. Still, it was raining, and the ditch was wet and deep.

"I am sorry we did not stop," I said, half to myself, "I can't help feeling that was a moth."

There is where youth, in all its impetuosity, helped me. If the girl had asked, "Shall I go back?" in all probability

I would have answered, "No, I must have been mistaken. Drive on!"

Instead, Molly-Cotton, who had straightened herself, and touched up her horse for a brisk entrance into town, said, "Well, we will just settle that 'feeling' right here!"

At a trot, she deftly cut a curve in the broad road and drove back. She drew close the edge of the ditch as we approached the lilies. As the horse stopped, what I had taken for a fallen lily bloom, suddenly opened to over five inches of gorgeous red-brown, canary-spotted wing sweep, and then closed again.

"It is a moth!" we gasped, with one breath.

Molly-Cotton cramped the wheel on my side of the carriage and started to step down. Then she dropped back to the seat.

"I am afraid," she said. "I don't want you to wade that ditch in the rain, but you never have had a red one, and if I bungle and let it escape, I never will forgive myself."

She swung the horse to the other side, and I climbed down. Gathering my skirts, I crossed the ditch as best I could, and reached the lily bed, but I was trembling until my knees wavered. I stepped between the lilies and the cornfield, leaned over breathlessly, and waited in the pelting rain, until the moth again raised its wings above its back. Then with a sweep learned in childhood, I had it.

While crossing the ditch, I noticed there were numbers of heavy yellow paper bags lying where people had thrown them when emptied of bananas and biscuits, on leaving town. They were too wet to be safe, but to carry the moth in my fingers would spoil it for a study, so I caught up and drained a big bag; carefully set my treasure inside, and handed it to Molly-Cotton. If you consider the word 'treasure' too strong to fit the case, offer me your biggest diamond, ruby, or emerald, in recompense for the privilege of striking this chapter, with its accompanying illustration, from my book, and learn what the answer will be.

When I entered the carriage and dried my face and hands, we peeped, marvelled, and exclaimed in wonder, for this was the most gorgeous moth of our collections. We hastened to Portland, where we secured a large box at a store. In order that it might not be dark and set the moth beating in flight, we copiously punctured it with as large holes as we dared, and bound the lid securely. On the way home we searched the lilies and roadside for a mile, but could find no trace of another moth. Indeed, it seemed a miracle that we had found this one late in August, for the time of their emergence is supposed to be from middle May to the end of June. Professor Rowley assures me that in rare instances a moth will emerge from a case or cocoon two seasons old, and finding this one, and the Luna, prove it is well for nature students to be watchful from May until October. Because these things happened to me in person, I made bold to introduce the capture of a late moth into the experience of Edith Carr in the last chapter of "A Girl of the Limberlost." I am pointing out some of these occurrences as I come to them, in order that you may see how closely I keep to life and truth, even in books exploited as fiction. There may be such incidents that are pure imagination incorporated; but as I write I can recall no instance similar to this, in any book of mine, that is not personal experience, or that did not happen to other people within my knowledge, or was not told me by some one whose word I consider unquestionable; allowing very little material indeed, on the last provision.

There is one other possibility to account for the moth at this time. Beyond all question the gorgeous creature is of tropical origin. It has made its way north from South or Central America. It occurs more frequently in Florida and Georgia than with us, and there it is known to have been double brooded; so standing on the records of professional lepidopterists, that gives rise to grounds for the possibility that in some of our long, almost tropical Indiana summers, Regalis may be double brooded with us. At any rate, many people saw the living moth in my possession on this date. In fact, I am prepared to furnish abundant proof of every statement contained in this chapter; while at the same time admitting that it reads like the veriest fairy tale 'ever thought or wondered.'

The storm had passed and the light was fine, so we posed the moth before the camera several times. It was nervous business, for he was becoming restless, and every instant I expected him to fly, but of course we kept him guarded.

There was no hope of a female that late date, so the next step was to copy his colours and markings as exactly as possible. He was the gaudiest moth of my experience, and his name seemed to suit rarely well. Citheroma—a Greek poet, and Regalis—regal. He was truly royal and enough to inspire poetry in a man of any nation. His face-was orange-brown, of so bright a shade that any one at a glance would have called it red. His eyes were small for his size, and his antennae long, fine, and pressed against the face so closely it had to be carefully scrutinized to see them. A band of bright canary-yellow arched above them, his thorax was covered above with long silky, orange-brown hairs, and striped lengthwise with the same yellow. His abdomen was the longest and slenderest I had seen, elegantly curved like a vase, and reaching a quarter of an inch beyond the back wings, which is unusual. It was thickly covered with long hair, and faintly lined at the segments with yellow. The claspers were very sharp, prominent brown hooks. His sides were dotted with alternating red and orangebrown spots, and his thorax beneath, yellow. The under side of the abdomen was yellow, strongly shaded with orange-brown. His legs and feet were the same.

His fore-wings were a silvery lead colour, each vein covered with a stripe of orange-brown three times its width. The costa began in lead colour, and at half its extent shaded into orange-brown. Each front wing had six yellow spots, and a seventh faintly showing. Half an inch from the apex of the wings, and against the costa, lay the first and second spots, oblong in shape, and wide enough to cover the space between veins. The third was a tiny dot next the second. The hint of one crossed the next vein, and the other three formed a triangle; one lay at the costa about three-quarters of an inch from the base, the second at the same distance from the base at the back edge of the wing, and the third formed the apex, and fell in the middle, on the fifth space between veins, counting from either edge. These were almost perfectly round. The back wings were very hairy, of a deep orange-brown at the base, shading to lighter tones of the same colour at the edge, and faintly clouded in two patches with yellow.

Underneath the fore-wings were yellow at the base, and lead colour the remainder of their length. The veins had the orange-red outlining, and the two large yellow dots at the costa showed through as well as the small one beside them. Then came another little yellow dot of the same size, that did not show on the upper side, and then four larger round spots between each vein. Two of them showed in the triangle on the upper side full size, and the two between could be seen in the merest speck, if looked for very closely.

The back wings underneath were yellow three-fourths of their length, then next the abdomen began a quarter of an inch wide band of orange-brown, that crossed the wing to the third vein from the outer edge, and there shaded into lead colour, and covered the space to the margin. The remainder of the wing below this band was a lighter shade of yellow than above it. From tip to tip he measured five and a half inches, and from head to point of abdomen a little over two.

While I was talking Regalis, and delighted over finding so late in the season the only one I lacked to complete my studies of every important species, Arthur Fensler brought me a large Regalis caterpillar, full fed, and in the last stages of the two days of exercise that every caterpillar seems to take before going into the pupa state. It was late in the evening, so I put the big fellow in a covered bucket of soft earth from the garden, planning to take his picture the coming day. Before morning he had burrowed into the earth from sight, and was pupating, so there was great risk in disturbing him. I was afraid there were insects in the earth that would harm him, as care had not been taken to bake it, as should have been done.

A day later Willis Glendenning brought me another Regalis caterpillar. I made two pictures of it, although transformation to the pupa stage was so far advanced that it was only half length, and had a shrivelled appearance like the one I once threw away. I was disgusted with the picture at the time, but now I feel that it is very important in the history of transformation from caterpillar to pupa, and I am glad to have it.

Two days later, Andrew Idlewine, a friend to my work, came to the Deacon with a box. He said that he thought maybe I would like to take a picture of the fellow inside, and if I did, he wanted a copy; and he wished he knew what the name of it was. He had found it on a butternut tree, and used great care in taking it lest it 'horn' him. He was horrified when the Deacon picked it up, and demonstrated how harmless it was. This is difficult to believe, but it was a third Regalis and came into my possession at night again. My only consolation was that it was feeding, and would not pupate until I could make a picture. This one was six inches from tip to tip, the largest caterpillar I ever saw; a beautiful blue-green colour, with legs of tan marked with black, each segment having four small sharp horns on top, and on the sides an oblique dash of pale blue. The head bore ten horns. Four of these were large, an inch in length, coloured tan at the base, black at the tip. The foremost pair of this formidable array turned front over the face, all the others back, and the outside six of the ten were not quite the length of the largest ones.

The first caterpillar had measured five inches, and the next one three, but it was transforming. Whether the others were males and this a female, or whether it was only that it had grown under favourable conditions, I could not tell. It was differently marked on the sides, and in every way larger, and brighter than the others, and had not finished feeding. Knowing that it was called the 'horned hickory devil' at times, hickory and walnut leaves were placed in its box, and it evinced a decided preference for the hickory. As long as it ate and seemed a trifle larger it was fed. The day it walked over fresh leaves and began the preliminary travel, it was placed on some hickory sprouts around an old stump, and exposures made on it, or rather on the places it had been, for it was extremely restless and difficult to handle. Two plates were spoiled for me by my subject walking out of focus as I snapped, but twice it was caught broadside in good position.

While I was working with this caterpillar, there came one of my clearest cases of things that 'thrust themselves upon me.' I would have preferred to concentrate all my attention on the caterpillar, for it was worth while; but in the midst of my work a katydid deliberately walked down the stump, and stopped squarely before the lens to wash her face and make her toilet. She was on the side of the stump, and so clearly outlined by the lens that I could see her long wavering antennae on the ground glass, and of course she took two plates before she resumed her travels. I long had wanted a katydid for an illustration. I got that one merely by using what was before me. All I did was to swing the lens about six inches, and shift the focus slightly, to secure two good exposures of her in fine positions. My caterpillar almost escaped while I worked, for it had put in the time climbing to the ground, and was a yard away hurrying across the grass at a lively pace.

Two days later it stopped travelling, and pupated on the top of the now hardened earth in the bucket that contained the other two. It was the largest of the pupae when it emerged, a big shining greenish brown thing flattened and seeming as if it had been varnished. On the thin pupa case the wing shields and outlines of the head and different parts of the body could be seen. Then a pan of sand was baked, and a box with a glass cover was filled. I laid the pupa on top of the sand, and then dug up the first one, as I was afraid of the earth in which it lay. The case was sound, and in fine condition. All of these pupae lived and seemed perfect. Narrow antennae and abdominal formation marked the big one a female, while broader antlers and the clearly outlined 'claspers' proved the smaller ones males. A little sphagnum moss, that was dampened slightly every few days, was kept around them. The one that entered the ground had pushed the earth from it on all, sides at a depth of three inches, and hollowed an oval space the size of a medium hen egg, in which the pupa lay, but there was no trace of its cast skin. Those that pupated on the ground had left their skins at the thorax, and lay two inches from them. The horns came off with the skin, and the lining of the segments and the covering of the feet showed. At first the cast skins were green, but they soon turned a dirty grey, and the horns blackened.

So from having no personal experience at all with our rarest moth, inside a few days of latter August and early September, weeks after hope had been abandoned for the season, I found myself with several as fine studies of the male as I could make, one of an immense caterpillar at maturity, one half-transformed to the moth, and three fine pupa cases. Besides, I had every reason to hope that in the spring I could secure eggs and a likeness of a female to complete my illustration. Call this luck, fairy magic, what you will, I admit it sounds too good to be true; but it is.

All winter these three fine Regalis pupa cases were watched solicitously, as well as my twin Cecropias, some Polyphemus, and several ground cocoons so spun on limbs and among debris that it was not easy to decide whether they were Polyphemus or Luna. When spring came, and the Cecropias emerged at the same time, I took heart, for I admit I was praying for a pair of Regalis moths from those pupa cases in order that a female, a history of their emergence, and their eggs, might be added to the completion of this chapter. In the beginning it was my plan to use the caterpillars, and give the entire history of one spinning, and one burrowing moth. My Cecropia records were complete; I could add the twin series for good measure for the cocoon moth; now if only a pair would come from these pupa cases, I would have what I wanted to compile the history of a ground moth.

Until the emergence of the Cecropias, my cocoons and pupa cases were kept on my dresser. Now I moved the box to a chair beside my bed. That was a lucky thought, for the first moth appeared at midnight, from Mr. Idlewine's case. She pushed the wing shields away with her feet, and passed through the opening. She was three and one-half inches LONG, with a big pursy abdomen, and wings the size of my thumbnail. I was anxious for a picture of her all damp and undeveloped, beside the broken pupa case; but I was so fearful of spoiling my series I dared not touch, or try to reproduce her. The head and wings only seemed damp, but the abdomen was quite wet, and the case contained a quantity of liquid, undoubtedly ejected for the purpose of facilitating exit. When you next examine a pupa, study the closeness with which the case fits antennae, eyes, feet, wings, head, thorax, and abdominal rings and you will see that it would be impossible for the moth to separate from the case and leave it with down intact, if it were dry.

Immediately the moth began racing around energetically, and flapping those tiny wings until the sound awakened the Deacon in the adjoining room. After a few minutes of exercise, it seemed in danger of injuring the other cases, so it was transferred to the dresser, where it climbed to the lid of a trinket case, and clinging with the feet, the wings hanging, development began. There was no noticeable change in the head and shoulders, save that the down grew fluffier as it dried. The abdomen seemed to draw up, and became more compact. No one can comprehend the story of the wings unless they have seen them develop.

At twelve o'clock and five minutes, they measured two-thirds of an inch from the base of the costa to the tip. At twelve fifteen they were an inch and a quarter. At half-past twelve they were two inches. At twelve forty-five they were two and a half; and at one o'clock they were three inches. At complete expansion this moth measured six and a half inches strong (sic!), and this full sweep was developed in one hour and ten minutes. To see those large brilliantly-coloured wings droop, widen, and develop their markings, seemed little short of a miracle.

The history of the following days is painful. I not only wanted a series of this moth as I wanted nothing else concerning the book, but with the riches of three fine pupa cases of it on hand, I had promised Professor Rowley eggs from which to obtain its history for himself. I had taxed Mr. Rowley's time and patience as an expert lepidopterist, to read my text, and examine my illustration; and I hoped in a small way to repay his kindness by sending him a box of fertile Regalis eggs.

The other pupa cases were healthful and lively, but the moths would not emerge. I coaxed them in the warmth of closed palms—I even laid them on dampened moss in the sun in the hope of softening the cases, and driving the moths out with the heat, but to no avail. They would not come forth.

I had made my studies of the big moth, when she was fully developed; but to my despair, she was depositing worthless eggs over the inside of my screen door.

Four days later, the egg-laying period over, the female, stupid and almost gone, a fine male emerged, and the following day another. I placed some of the sand from the bottom of the box on a brush tray, and put these two cases on it, and set a focused camera in readiness, so that I got a side view of a moth just as it emerged, and one facing front when about ready to cling for wing expansion. The history of their appearance, was similar to that of the female, only they were smaller, and of much brighter. colour. The next morning I wrote Professor Rowley of my regrets at being unable to send the eggs as I had hoped.

At noon I came home from half a day in the fields, to find Raymond sitting on the Cabin steps with a big box. That box contained a perfect pair of mated Regalis moths. This was positively the last appearance of the fairies.

Raymond had seen these moths clinging to the under side of a rail while riding. He at once dismounted, coaxed them on a twig, and covering them with his hat, he weighted the brim with stones. Then he rode to the nearest farmhouse for a box, and brought the pair safely to me. Several beautiful studies of them were made, into one of which I also introduced my last moth to emerge, in order to show the males in two different positions.

The date was June tenth. The next day the female began egg placing. A large box was lined with corrugated paper, so that she could find easy footing, and after she had deposited many eggs on this, fearing some element in it might not be healthful for them, I substituted hickory leaves.

Then the happy time began. Soon there were heaps of pearly pale yellow eggs piled in pyramids on the leaves, and I made a study of them. Then I gently lifted a leaf, carried it outdoors and, in full light, reproduced the female in the position in which she deposited her eggs, even in the act of placing them. Of course, Molly-Cotton stood beside with a net in one hand to guard, and an umbrella in the other to shade the moth, except at the instant of exposure; but she made no movement indicative of flight.

I made every study of interest of which I could think. Then I packed and mailed Professor Rowley about two hundred fine fertile eggs, with all scientific data. I only kept about one dozen, as I could think of nothing more to record of this moth except the fact that I had raised its caterpillar. As I explained in the first chapter, from information found in a work on moths supposed to be scientific and accurate, I depended on these caterpillars to emerge in sixteen days. The season was unusually rainy and unfavourable for field work, and I had a large contract on hand for outdoor stuff. I was so extremely busy, I was glad to box the eggs, and put them out of mind until the twenty-seventh. By the merest chance I handled the box on the twentyfourth, and found six caterpillars starved to death, two more feeble, and four that seemed lively. One of these was bitten by some insect that clung to a leaf placed in their box for food, in spite of the fact that all leaves were carefully washed. One died from causes unknown. One stuck in pupation, and moulded in its skin. Three went through the succession of moults and feeding periods in fine shape, and the first week in September transformed into shiny pupa cases, not one of which was nearly as large as that of the caterpillar brought to me by Mr. Idlewine. I fed these caterpillars on black walnut leaves, as they ate them in preference to hickory.

I am slightly troubled about this moth. In Packard's "Guide to the Study of Moths", he writes: "Citheronia Regalis expands five to six inches, and its fore-wings are olive coloured, spotted with yellow and veined with broad red lines, while the hind wings are orange-red, spotted with olive, green, and yellow."

He describes two other species. Citheronia Mexicana, a tropical moth that has drifted as far north as Mexico. It is quite similar to Regalis, "having more orange and less red," but it is not recorded as having been found within a thousand miles of my locality. A third small species, Citheronia sepulcralis, expands only a little over three inches, is purple-brown with yellow spots; and is a rare Atlantic Coast species having been found once in Massachusetts, oftener in Georgia, never west of Pennsylvania.

This eliminates them as possible Limberlost species. Professor Rowley raised this moth from the eggs I sent him.

The trouble is this: Packard describes the fore-wings as 'olive,' the hind as 'olive, and green.' Holland makes no reference to colour, but on plate X, figure three, page eighty-seven, he reproduces Regalis with fore-wings of olive-green, the remainder of the colour as I describe and paint, only lighter. In all the Regalis moths I have handled, raised, studied minutely, painted, and photographed, there never has been tinge or shade of GREEN. Not the slightest trace of it! Each moth, male and female, has had a basic colour of pure lead or steel grey. White tinged with the proper proportions of black and blue gives the only colour that will exactly match it. I have visited my specimen case since writing the preceding. I find there the bodies of four Regalis moths, saved after their decline. One is four years old, one three, the others two, all have been exposed to daylight for that length of time. The yellows are slightly faded, the reds very much degraded, the greys a half lighter than when fresh; but showing to-day a pure, clear grey.

What troubles me is whether Regalis of the Limberlost is grey, where others are green; or whether I am colour blind or these men. Referring to other writers, I am growing 'leery' of the word 'Authority'; half of what was written fifty years ago along almost any line you can mention, to-day stands disproved; all of us are merely seekers after the truth: so referring to other writers, I find the women of Massachusetts; who wrote "Caterpillars and Their Moths", and who in all probability have raised more different caterpillars for the purpose of securing life history than any other workers of our country, possibly of any, state that the front wings of Regalis have "stripes of lead colour between the veins of the wings," and "three or four lead-coloured stripes" on the back wings. The remainder of my description and colouring also agrees with theirs. If these men worked from museum or private collections, there is a possibility that chemicals used to kill, preserve, and protect the specimens from pests may have degraded the colours, and changed the grey to green. But to accept this as the explanation of the variance upsets all their colour values, so it must not be considered. This proves that there must be a Regalis that at times has olive-green stripes where mine are grey; but I never have seen one.

I think people need not fear planting trees on their premises that will be favourites with caterpillars, in the hope of luring exquisite te moths to become common with them. I have put out eggs, and released caterpillars near the Cabin, literally by the thousand, and never have been able to see the results by a single defoliated branch. Wrens, warblers, flycatchers, every small bird of the trees are exploring bark and scanning upper and under leaf surfaces for eggs and tiny caterpillars, and if they escape these, dozens of larger birds are waiting for the half-grown caterpillars, for in almost all instances these lack enough of the hairy coat of moss butterfly larvae to form any protection. Every season I watch my walnut trees to free them from the abominable 'tent' caterpillars; with the single exception of Halesidota Caryae, I never have had enough caterpillars of any species attack my foliage to be noticeable; and these in only one instance. If you care for moths you need not fear to encourage them; the birds will keep them within proper limits. If only one person enjoys this book one-tenth as much as I have loved the work of making it, then I am fully repaid.